SUBSTITUTE NATURAL GAS
MANUFACTURE AND PROPERTIES

Substitute Natural Gas

Manufacture and Properties

W. L. LOM

European Investment Bank, Luxembourg

and

A. F. WILLIAMS

Principal Scientist, Esso Research Centre,
Abingdon, Oxon, England

APPLIED SCIENCE PUBLISHERS LTD

LONDON

APPLIED SCIENCE PUBLISHERS LTD
RIPPLE ROAD, BARKING, ESSEX, ENGLAND

ISBN: 0 85334 671 2

WITH 26 ILLUSTRATIONS AND 51 TABLES

Printed in Great Britain by Galliard (Printers) Limited, Great Yarmouth

Foreword

Synthetic fuels generally are expected to become significant forms of energy supply before the end of this century. Substitute natural gas will be one form of these synthetics, and that is the subject of this book. Other forms will be lower heating value gases, which could be manufactured using processes described here, and a range of liquid products. They will be used to supplement and eventually largely replace supplies of fuels and feedstocks derived from natural gas and conventional crude oil. Coal, lignites through bituminous grades, will be the major resources for conversion to synthetics because of their abundance. Oil shale, tar sands and extra heavy oil deposits also will be important resources for synthetics.

Energy and the coming need for synthetic fuels have been dealt with at length in various media. Suffice it to say here that the need for synthetic fuels stems from (1) expectations of continued economic growth, (2) linkages between energy use and economic activity, (3) risks of dependence on imported oil and gas, (4) approaching resource limits on crude oil and natural gas supply, (5) much more distant resource limits on coal, oil shale, etc., and (6) inertias at energy end uses preventing large, short-term substitutions of electricity, coal, etc., for gas and liquid fuels. The future need for synthetics will be the net effect of these factors. However, none of them can be quantified accurately and forecasts of need, calculated as differences between large numbers, can vary widely.

The need problem becomes complicated further when one considers possible supply profiles for synthetics. To illustrate, let us assume first that 1·5 million barrels per day (MB/D) of synthetics are produced in 1985; perhaps the 1 MB/D considered to be the target level in the US and the rest elsewhere. Second, assume that synthetics production could grow fifteen percent annually. In 1995, production then could be 6 MB/D. Would that be enough? For reference, oil and gas consumption in the OECD countries in 1975 is about 55 MB/D oil equivalent and, if supplies were available, probably would be upwards from 90 MB/D in 1995.

Whatever profile of synthetics supply proves to be needed, getting any requires getting started. The first problem in getting started is posed by economics, and the second by technology availability, particularly for coal liquefaction, but also for advanced, high train capacity coal gasification processes. The economics problem is that levels of investment required per unit of synthetics capacity are higher than those experienced so far for conventional crude oil and natural gas production. Thus, prospective entrants into the synthetics industry need assurance of project viability by means of higher product prices or some kind of government incentive, guarantee or subsidy. This need will disappear eventually as conventional production becomes more and more difficult and capital intensive. Prospects for significant reduction in investment required per unit of synthetics capacity, however, are not great for the surface production synthetics processes now developing.

The problem of technology availability derives partly from the problem of economics. Until economic viability of synthetics technology is apparent, there is high risk associated with large expenditures for developing new technology. Also, development of new, complex processes requires time and technical success cannot be guaranteed. Development time for a new process will be affected not only by the complexity of the process but also by the number of steps in the development plan. Let us assume that a new process has operated successfully on a one ton/day pilot plant scale and that a commercial application would be sized at 15 000 tons/day. If precommercial development could be completed in one step by designing, building and operating a suitably sized pilot plant, say of several hundred tons/day capacity, time to full commercial operation from the 1 ton/day scale might be 8–10 years. Each additional development step in pilot or demonstration scale plants would increase development time and cost significantly. Any process now under development that will need to proceed through a demonstration plant step can hardly be ready for commercial use much earlier than 1985. Thus, unless the development pace for new synthetics technology is put on a crash basis, most capacity to be put in place by the early 1980s will utilise processes already proven. Such processes exist for producing liquids from oil shale, tar sands and heavy oil deposits as well as for gasifying coals.

The authors of this book have worked together for a number of years in a world gas technical service function at the Esso Research Centre in England, and thus have been well situated to keep abreast of the rapidly developing technology for SNG. In the recent past, they authored two additional books on the manufacture and properties of liquefied petroleum gas and liquefied natural gas. They also have published a number of papers dealing with the relationships between feedstock qualities and gasification processes.

In the present book, emphasis is given to processes for the manufacture of SNG from coal but SNG manufacture from petroleum feedstocks is also covered from both technical and economic angles. These subjects are supplemented by introductory chapters on gas and comments by the authors on other replacement energy sources. The book should provide all who are concerned about future energy supplies with a useful introduction to the SNG possibilities.

L. E. SWABB, JR.
Vice-President, Synthetic Fuels Research
Exxon Research and Engineering Company
Florham Park
New Jersey 07932
USA

Preface

The subject of the different forms of energy supplies for the future development of mankind has been raised on numerous occasions by many authors in the recent past. While only a few particularly clear minded scientists and engineers had expressed misgivings about the way modern technology consumed, and wasted, large amounts of energy to maintain a privileged fraction of the human race in the comfort to which it had become accustomed, the number of prophets of doom has proliferated since the last Arab–Israeli war and the subsequent drastic increase in oil prices. Suddenly we have all become energy conservation conscious.

This sudden conversion certainly did not take place too soon. There are many complex problems to be solved, in regard to both reduced wastage and new or modified energy sources, before we can claim to have satisfied the fuel requirements of the next few generations, not to speak of a permanent solution of the energy, food and raw materials deficiencies which the growing numbers of passengers on space-ship Earth must ultimately produce.

Fossilised hydrocarbons, i.e. lignite, coal, crude oil, natural gas, tar sands and oil shale, took vastly longer to develop in the Earth's crust than the rate at which they are being consumed at present. There is little doubt, therefore, that the first shortages will arise in this area and that substitution, supplementation and a high degree of economy in their consumption will be essential if we want these scarce materials to last us, at least until we have learned to live without them.

Of the fossil fuels it is natural gas that shows the first signs of approaching, if not total exhaustion, at least local shortages and there is a need to transfer large volumes from areas of local surplus to regions of high demand. Substitution and/or supplementation of natural gas would, therefore, appear to be one of the first, if not the most immediately important, replacement steps required in the field of energy technology.

The authors do not claim that such general philosophical considerations have led them to write the book. It would in fact be a little disingenuous to assert that we chose this subject because of humanity's more urgent need to replace natural gas rather than to supplement other fuels and raw materials. It had, however, become fairly clear that considerable technical effort in the United States, Europe and Japan had tended towards work on substitution of gaseous fuels over the last few years. New technologies were developing and older techniques were being modified to convert liquid and solid fuels, whose supplies were considered more reliable than those of natural gas, into a clean gaseous fuel of similar combustion characteristics to those of natural gas itself.

While it is difficult to assess the exact moment when a rapidly developing technology has reached a plateau and is therefore ripe for summarisation and critical review, it seemed to us that just such a situation had arisen in regard to manufacture of substitute natural gas (SNG). Although detailed improvements would, no doubt, continue to be made in gasification processes, and new raw material sources would continue to be tapped, the basic lines of conversion processes seemed to us to have been established, and in addition gasification routes for the most suitable hydrocarbon fuels now appear to be available.

In line with the rapidly developing SNG technology, the literature has grown apace in the last five years or so. It has been the authors' intention to chart a way through the mass of detail presently available by providing a logical breakdown of subject headings. What has emerged is this book in which the chapters deal respectively with natural gas shortages, a description of oil and coal based processes, underlying chemical reactions, feedstock supply and quality angles, interchangeability of product gas with the naturally occurring counterpart, comparative economics and, finally, a guess at the various ways in which energy may be made available, and consumed, in future years.

The book will serve, we think, to introduce the exciting SNG prospect to a wide variety of interests in the field of fuel technology and engineering on the one hand and those concerned with the gas industry and other energy supply organisations on the other.

Those wishing to delve more deeply into the detail of any particular topic will no doubt start with the references provided at the end of each chapter.

W. L. LOM
A. F. WILLIAMS

Contents

Acknowledgements

This book was prepared and largely written whilst both authors were working within the Exxon organisation. One of the authors (Walter Lowenstein-Lom) has since left and is now Energy Adviser to the European Investment Bank, Luxembourg. The authors are grateful to Exxon for permission to publish and to colleagues who over many years, perhaps unwittingly, have helped by wholeheartedly joining in stimulating discussion on matters affecting the world energy scene. It is to be noted, however, that the authors have written this book in a private capacity and that the views expressed therefore are entirely their own and not necessarily those of their respective employers.

Thanks are particularly due to Mme Menguy who typed most of the book and checked much of the material.

Finally, the authors pay tribute to their respective wives, Dinah and Lesley, who have long put up with what must have seemed to them a monopolising of time which could have been more properly devoted to their families.

Chapter 1

An Introduction to SNG

1.1 BRIEF HISTORY OF THE GAS INDUSTRY

The observation that a gaseous fuel[1,3] piped to individual households, industrial and commercial consumers, and ultimately even to large users of energy such as power stations was convenient, clean and could also be economical, was made some 150 years ago when the first city streets were lit by gas lamps. In the years that followed, distribution of fuel gas by pipes, originally for street lighting, subsequently for domestic illumination, then for cooking, baking and various commercial uses, and more recently for domestic heating, became widely established.

By the beginning of World War I practically all large and medium-sized cities in the temperate zones of the world and even many towns in tropical areas had extensive gas distribution networks that guaranteed the bulk of their population a constant supply of a gaseous fuel of constant quality.

It is worth noting that the gas in question, almost everywhere in the world, in the sense of being manufactured, was 'synthetic' and that coal, with a very few exceptions, was the raw material from which it was produced. Gasworks were built in all these cities in which bituminous coal was subjected to devolatilisation and partial thermal cracking in horizontal or vertical retorts to produce, on the one hand, a solid char or gas coke suitable mainly for firing in domestic enclosed stoves or central-heating boilers, on the other hand a combustible gas that, after suitable processing and purification, was an ideal fuel for lighting, cooking and space heating. Thus did coal gas containing some 20–30 % vol methane and about 50 % vol hydrogen (Table 1.1) become the foundation of the early 500 Btu/ft^3 town gas industry.

Clearly the cost of coal processing by 'carbonisation' was not negligible and the two fuel products were somewhat more expensive

1

TABLE 1.1
Characteristics of Various Utility and Industrial Gases

	Water Gas		Producer gas from coke	Blast furnace gas	Coke-oven gas	Town Gas			Oil gas	Natural Gas[b]	LPG	
	Blue	Carburetted				H.R.	V.R.	L.T.			C_3	C_4
Composition (% vol.)												
Oxygen	—	0·4	—	—	0·5	0·5	0·8	0·2	—	—	—	—
Carbon dioxide	4·5	5·6	5·0	11	2·0	2·0	4·0	4·5	15·0	(0·5)	—	—
Carbon monoxide	41·0	30·5	29·0	27	7·5	8·0	18·0	8·3	2·5	—	—	—
Hydrogen	50·0	37·0	11·0	—	55·0	52·0	50·0	29·1	65·0	—	—	—
Methane	0·5	14·0	0·5	—	27·0	30·0	20·0	49·1	10·0	80–99	—	—
Higher hydrocarbons[a]	—	7·0	—	—	2·5	3·5	2·0	3·8	7·5	3–20	(95)[c]	(95)[c]
Nitrogen	4·0	5·5	54·5	62	5·5	4·0	5·2	5·0	—	0·3–14·3	—	—
Relative density approx., 15°C, 1 atm (air = 1)	—	—	0·9	—	0·38	0·42	0·48	0·5	0·5	0·55–0·65	1·5	2·0
Heating value Btu/ft³ (60°F, 30 in Hg, sat'd)	290	500	150	190	525	560	500	500	500	(1000)[c]	2 500	3 200
kcal/m³ (0°C, 760 mm Hg, dry)	2 760	4 750	1 425	1 810	5 000	5 320	4 750	4 750	4 750	9 500	23 700	30 300

H.R.   Horizontal retorts (coal)
V.R.   Vertical retorts (coal)
L.T.   Low temperature coal-fired retorts
[a]   Includes unsaturated hydrocarbons
[b]   This book, Chapter 2
[c]   Nominal
C_3 Propane
C_4 Butane

than the raw material, coal. Hence gas would be used mainly for specific applications where its superiority as a fuel was undisputed, such as lighting and domestic cooking; however, in both these fields relatively cheap and efficient electricity soon became a formidable competitor. Furthermore, outlets for gasworks coke, the inevitable by-product of carbonisation, were limited since both coal and oil tended to compete for the domestic heating market, where even its property of smoke-free combustion did not guarantee it sufficient outlets.

One of the solutions offered by the gas industry was to gasify the coke, for instance by means of steam in a gas generator, where alternative blasts of air and steam would provide heat energy and a supply of 'blue' water gas (carbon monoxide and hydrogen) respectively. Both water gas (also known as carburetted water gas when enriched with cracked oil gas) and gas made by complete gasification in special generators were used in efforts to reduce the cost and increase the availability of coal gas.[2]

Another source of coal gas in some countries was coke-oven gas, an unavoidable by-product of the process of heating bituminous coals in coke ovens to manufacture metallurgical coke for iron and steel making purposes. Attempts were also made to combine the generation of utility gas with the process of gasifying coal to make such industrial chemicals as ammonia and methanol from intermediate 'synthesis gas' (a mixture of carbon monoxide and hydrogen).

Ultimate success of these efforts did not materialise, mainly for two reasons; the price of coal, particularly in the period after World War II in many parts of the globe and particularly in Europe, rose to a level well above that of imported liquid petroleum fuels; in addition, discoveries of predominantly methane natural gas led to the substitution of the latter for coal gas in many existing distribution grids, as for example in Southern France and Italy.

The gas industry, particularly in Europe, now found itself in a strange dilemma. Coal gas had become expensive, yet a very large number of consumers with an even larger number of appliances were connected to various gas grids and were, to all intents and purposes, captive customers. However, as a result of rising gas prices, consumption became at least stagnant, sometimes even reduced, and the incidence of rising fixed costs on gas price became more significant.

Clearly wherever cheap natural gas (9–$10\,000\,kcal/m^3$) was obtainable one would try to replace the old coal-based town gas supply ($4500\,kcal/m^3$) although this meant existing burners and appliances had to be modified or replaced to allow for the different combustion characteristics of the new supply. Gas appliance conversion, in fact, became an important industry and, as will be mentioned later, many millions of gas customers all over the world have since been 'converted', albeit at high cost to the gas supplier.[5]

However, before natural gas had reached its present dominant position in world-wide gas supplies, another generation of town gas manufacturing processes was to be developed using cheaper petroleum fuels as feedstocks.

The first series of oil-based processes partially oxidised fuel oil in cyclic systems of gas generation and catalyst reactivation to produce once again $500\,Btu/ft^3$ town gas (Hall, Segas, Dayton processes). Later these were to be largely supplanted by less complicated and cheaper processes of catalytic steam reforming of light virgin naphtha, again employing cyclic systems for small send-outs (Onia Gegi, Micro Simplex) or a continuous version requiring desulphurised naphtha for larger send-outs (ICI town gas or '500' process, Haldor Topsøe).[3] Thus were several hundreds of naphtha plants installed, principally in the United Kingdom but also in Japan, France, Germany, Italy and the United States.

Again there is little doubt that this approach would have been acceptable both economically and technologically, and the gas industry might have continued manufacturing and distributing oil-based gases in all those countries where natural gas could not be easily introduced. However, another set of historical accidents occurred: at least three of the main industrial concentrations in the world found that natural gas was, in fact, present in economic quantities in the subsoil within pipelining distance from their industries. In the United States and Canada, Russia and, more recently, Europe, natural gas has become both the most widely used gaseous fuel and also the quality standard against which all other fuel gases have to be measured.

Particularly in North America, partly because the distribution of coal gas had never become widely established, natural gas made rapid progress. It was piped in large volumes, first from the Appalachians, later from further afield into the huge conurbations on the Eastern

Seaboard. Other pipelines from the Deep South and Texas followed the Mississippi and supplied the industries in the Great Lakes area. Imports of Canadian natural gas reached both the Midwest and the Pacific coast. Finally the smaller industrial centres were linked to nearby fields and to long distance pipelines that ultimately provided practically complete natural gas coverage of the entire US and Canada.[2]

Similarly in Russia, the Moscow area was first supplied with natural gas from wells in the Northern Ukraine; later, connections were established with the oilfields between the Black Sea and the Caspian and supplies were directed both to the North and to the industrial areas in the Southern Ukraine; next wells in the Urals were linked to the system, and most recently some of the very large gas reserves in Siberia have been tapped. Gas is now supplied not only to Russian industrial and domestic customers but also to all the Eastern European satellites and, in increasing volumes, even to Western Europe.

The situation in the EEC countries is a very special one. Natural gas was found in reasonable quantities soon after World War II in Italy, Germany and France. Owing to somewhat improvident utilisation, reserves were, however, rapidly exhausted and only the discovery of a very large deposit in Groningen, Holland, postponed the eventual day of reckoning.[4] Even more recently (1965), in the UK, substantial quantities were found in the Southern parts of the North Sea which, for a period, seemed likely to meet most of the UK requirements.[6] However, it quickly became evident that peak production rates would be levelled off in the mid 1980s and in order to meet growing demand of the area, it will be necessary not only to produce more gas in Northern North Sea waters but also to import large volumes of pipeline gas from Norway and possibly Russia as well as liquefied gas (LNG) from Middle East countries.

The only major industrial nation that has not so far changed its basic gas supply from coal gas to natural gas is Japan, although even there certain parts of the country are being converted to LNG imported from Alaska and Indonesia.

Many other, slightly less important conurbations such as Melbourne, Perth, Adelaide and Sydney in Australia, Auckland and Christchurch in New Zealand, Buenos Aires and Mexico City will all be receiving natural gas from local mainland or offshore gas fields. In all these distribution areas old gas appliances have been converted and new appliances are being manufactured to natural gas rather than coal gas standards.

It is, therefore, the more distressing to note that in many of these towns and countries newly discovered natural gas reserves are predicted to be unable to keep up with consumption within the next ten years or so, a picture that has been foreshadowed by gas supply patterns in Italy, France and the Netherlands. It is to the question of coping with shortfalls in natural gas availability that the bulk of this book is now addressed.

Before leaving the discussion on historical gas supplies, it is however of interest to refer the reader to the composition of the various coal and oil gases previously described and to indicate the wide differences in composition between tributary streams to town gas supplies as distinct from the more uniform natural gas supply picture (Tables 1.1 and 2.3).

1.2   THE IMPENDING NATURAL GAS SHORTAGE

The reasons for the imbalance between gas consumption and new gas discoveries are manifold. The search for gas, in view of restrictions imposed on gas prices by Governments, nationalised distribution companies and other monopoly buyers, had always taken second place to oil prospecting in most areas. Natural gas, co-produced with oil, is becoming less important owing to better production techniques, re-injection of gas for secondary oil recovery, and also the remote location of most recent large oil finds. Controlled low gas prices, furthermore, have provided an incentive to use natural gas for many applications over and above those justified by its undoubted technical advantages. Finally, our recent great concern about pollution and the ecological effects of industrial activities have provided a further reason for burning clean gaseous fuels.

For all these reasons first the United States and subsequently most European countries have predicted imminent shortages of natural gas which will require either drastic cuts in consumption or a substantial effort to find additional natural gas or to develop a suitable replacement.

Little need be said about the search for new natural gas deposits. Prospecting is in progress in remote locations such as Alaska, the

Canadian North, Eastern Siberia, Algeria, Iran, Malaysia and the Northern North Sea. Laying of pipelines from newly discovered fields in these areas (and substantial finds have already occurred) will take many years and require very large investments. Political considerations often militate against pipelines through third countries such as the much discussed gas lines from Iran and Algeria to Western Europe, and gas deliveries from one country to another are often limited by concern about future reserves.

Less politically sensitive as far as pipeline way-leaves are concerned, but more so in other respects, are LNG schemes. Here gas is piped from the field to the seashore, liquefied by compression and refrigeration and shipped in specially designed vessels to the country of destination. On arrival the liquid is pumped ashore and regasified before being injected into and distributed through the pipeline system of the distribution company. Investment in liquefaction plant and ships is very large and particularly the former, if owned or financed by the purchaser of the gas, can become a hostage in the gas supplier's hand.

Additional gas imports by pipeline from the remaining countries with a prospective surplus such as Russia, Norway and some of the Middle East countries is, clearly, also a possible means of mitigating the expected shortage. Additional supply contracts have, in fact, recently been concluded between Russia and Germany, Russia and Austria, Russia and Finland/Sweden and Norway and Sweden. All these agreements, however, have one feature in common: natural gas prices are very much higher than those negotiated originally.

The uncertainty overshadowing the whole issue of whether or not future gas supplies can meet the expected demand is illustrated by estimates relating to the US supply/demand picture over the next twenty years or so (Table 1.2).

Whilst the demand is expected to rise from 36 to 55 trillion ft^3/yr, the supply available from US production, including the new Alaskan finds and allowable LNG imports, is forecasted to remain steady at about 24 trillion ft^3/yr. The inevitability of a widening gap can therefore be expected to have a number of possible consequences, amongst which are the following: a reduction in total gas consumption: conversion of some traditional gas markets to other fuels; change in fiscal laws to allow additional LNG imports: make-up by gas manufactured to quality standards that provide for full interchangeability between 'natural' and 'synthetic' sources.

TABLE 1.2
Estimate of US Needs for Gas (trillion 10^{12} ft^3/yr)

Year	Total requirement	Available[a]	Need[b]
1980	35·7	22·7	13·0
1985	40·8	22·8	18·0
1990	47·3	23·7	23·6
1995	55·2	24·9	30·3

1 ft^3 = 0·028 32 m^3

[a] Expected to derive from own natural gas production, including Alaskan supplies.
[b] Gap indicating scope for SNG manufacture or LNG imports, etc.

Source: US Department of the Interior, US Bureau of Mines (1971). Mineral Industry Surveys. *Natural Gas Production and Consumption*, 1970.
Linden, H. R. (1972). Paper No. SPE 3581 to Society of Petroleum Engineers of AIME, New Orleans.
Gas Industry Committee, Future Requirements Agency, University of Denver (1971). *Future Gas Requirements of the USA, Vol. 4.*

It is to this concept of a 'synthetic' replacement gas that we devote the next section of this introductory chapter.

1.3   SNG

For all these reasons the replacement, partial substitution or at least supplementation of an existing natural gas supply with a synthetic gas made from a raw material that is either cheaper or at least more easily available than natural gas has become of considerable interest. A large number of process routes starting from various raw materials and ending up with a suitable gaseous fuel have been proposed and will be discussed in this book.

The gaseous fuels referred to have been given the collective name of SNG and the first of the three letters has received a number of interpretations. Originally SNG stood for synthetic natural gas, until some brave defender of pure English and logical thinking pointed out that what was natural could not very well be synthetic at the same time. However, the acrostic had by then become established and any other adjective or descriptive term had to have the initial S. 'Substitute' natural gas appears to have become the preferred interpretation of the three initials, but 'supplemental' natural gas would probably be the more logical one.

Logical because if one wished to substitute some other gas source permanently and irrevocably for an existing natural gas supply one would be free to choose, within certain limits, whatever type or quality of gas one wished to use. On the other hand if the replacement was to be used only temporarily or in admixture with the existing gas supply then clearly its properties, specifically in regard to combustion, would have to be precisely defined. In other words, supplemental gas would have to be fully compatible with natural gas as distributed at present. Most of the projects for the manufacture of large volumes of gas from liquid petroleum or solid fuels and other raw materials aim at producing a gas that is fully interchangeable with present supplies, *i.e.* by our definition a 'supplemental' gas. In some instances, which are relatively rare at present but may become more frequent in future, existing gas supplies will disappear completely and will have to be replaced by a new source of gas; by our definition this would be a substitute gas.

Both types of gases are synthetic in the sense that they require chemical processing and cannot simply be collected. They differ, however, from other synthetics in that their chemical composition, as a rule, is extremely close to, if not actually identical with, that of natural gas after the latter has been cooled, dried and purified for pipeline distribution. The composition of natural gases is in fact so simple, chemically, that replacements which not only have the right properties but are in fact nearly identical in composition can be produced without undue technical effort or excessive investment.

The exception to this rule is hydrogen, a gas which may at some future date replace the dwindling supplies of natural gas from fossil sources. While the purpose of gasifying fossil fuels will always be to produce methane, the main component of all types of natural gas, hydrogen could, in the absence of hydrocarbon fuels, become an acceptable replacement for natural gas with a number of additional valuable characteristics.

1.4 ECOLOGICAL CONSIDERATIONS

Natural gases contain, apart from methane, small quantities of the other low boiling, volatile hydrocarbons and certain trace impurities, which are as a rule removed before the gas is distributed. They are, therefore, extremely clean fuels that can be burnt without giving rise to any significant form of pollution. In contrast, solid and to some extent liquid fuels, when burnt, emit sulphur oxides, partially oxidised hydrocarbons, carbon monoxide, soot or other solid organic compounds and inorganic fly ash. The conversion of liquid or solid fuels into gases provides an opportunity to purify the fuel before distribution and combustion, and hence to reduce or eliminate the consequent atmospheric pollution. The gasification of other fossil fuels thus becomes desirable for the following reasons:

it facilitates handling and transportation not only of solids but also of liquids since both types of fuel are rarely piped in the small quantities required by individual consumers,
it permits precise control to be exerted over the processes of mixing and chemical reaction that occur during combustion,
it reduces pollution of the atmosphere by combustion gases and the possibility of ground or water pollution due to the spillage of non-gaseous fuels.

It follows that a decision to manufacture SNG can be the result of ecological considerations as well as the simple replacement of a material in diminishing supply by a substitute, the ample availability of which is assumed, or at least is less in doubt, than that of the original. Thus the recent decisions to gasify large quantities of coal in the South Western states of the USA are due both to the need to supplement natural gas, the demand for which is exceeding available supplies, and the desire to avoid the contamination of the atmosphere that would result from the burning of these bituminous and sub-bituminous coals.

It could be argued that the latter aim could be achieved by other means such as purification of the solid (or liquid) fuel or of the stack gases. Moreover, to avoid pollution one would not necessarily have to gasify the coal to produce SNG, and any other gaseous conversion would have been acceptable in that respect. It seems, however, and this is supported by the large number of projects that are now in the planning stage, that the SNG route is hardly more complex than other gasification systems, and we may find that SNG will be produced, whether to supplement or replace natural gas supplies, or solely for ecological reasons.

1.5   ROUTES TO SNG

The technology and cost of converting other fuels into gases interchangeable with natural gas will vary enormously and will depend largely on the properties and therefore the ease of conversion of the feedstock in question. While it is possible to make a good quality substitute from almost any fossil fuel such as coal or crude oil, or from any hydrocarbon fraction derived from these raw materials, both the complexity of the process and the processing cost will be much reduced if the fuel is of low molecular weight and simple chemical composition. Light hydrocarbons such as LPG, naphtha, gas condensate or jet fuel can be gasified relatively easily by interacting them with steam under suitable conditions. Heavier distillates will not easily react under such conditions, though certain claims have been made to that effect, and will as a rule require free hydrogen, prepared in a subsidiary unit, to initiate the gasification process.

Hydrogen as a gasifying agent can be used even further up the scale of feedstock complexity to gasify crude oil, residual fuels or coal but, during this process, reaction conditions tend to become so severe that it may be preferable to partially oxidise the raw material first. Hence, the three main process routes involved in the gasification of fossil fuels, each associated with a typical feedstock, are:

steam reforming of light distillates;
hydrogasification of middle distillates and residual fuels;
partial oxidation of residual fuels or coal.

1.6   ORGANISATION OF SUBJECT MATTER

In the following chapters it is proposed to develop the themes touched upon in this section. In the next chapter dealing with the availability, geographical distribution and properties of natural gases it is proposed to provide a rationale for the manufacture of substitutes from fossil fuels other than natural gas. Chapter 3 discusses gas combustion characteristics and gaseous interchangeability, a problem of great significance in regard to supplementation and substitution of existing supplies. The raw materials available for conversion into SNG and their properties and characteristics are the subject matter of Chapter 4. In Chapter 5 the principles of the technologies involved in gasification are described along with the thermochemistry, thermodynamics and kinetics of reforming, hydrogenation, partial oxidation, methanation and other reactions.

Chapters 6, 7 and 8 provide a somewhat more detailed approach to the three types of chemical processing, namely steam reforming, hydrogasification and partial oxidation, and the technologies involved in the gasification of liquid fuels ranging in volatility from LPG/naphtha to heavy fuel oil and crude oil itself.

The complex subject of solid fuel gasification and of the conversion of coal into SNG is treated in Chapter 9 which reviews the technologies of the new processes that are being developed, particularly in the United States. Chapter 10 deals solely with the technological aspects of methanating carbon oxides, a most important processing step in all conversions of liquid and solid fuels into SNG. Economic considerations are brought together in Chapter 11 and the book ends with a chapter that tries to look into the future and to analyse today's scientific and technical developments that may have a bearing on gasification trends in the years ahead, *e.g.* nuclear energy-assisted gasification processes, alternatives to the manufacture of supplemental gas, such as LNG or methanol imports, and finally to the replacement of present supplies of natural gas by an economy based on hydrogen or lean gases confined in industrial gas grids.

REFERENCES

1. Ministry of Fuel and Power (1944). *The Efficient Use of Fuel*, HMSO, London.
2. Pacific Coast Gas Association (1934). *Gas Engineer's Handbook*, McGraw-Hill, New York.
3. Ward, E. R. (1965). *Gasmaking*, 2nd Ed., British Petroleum Co., London.
4. Organisation for economic cooperation and development (1969). *Impact of Natural Gas on the Consumption of Energy*, OECD Paris.
5. Doherty, C. H. (1970). *New Gas for Old, the Story of Natural Gas*, Clifton Books, London.
6. Scott Wilson, D. (1969). *The Modern Gas Industry*, Edward Arnold, London.

Chapter 2

Review of Natural Gases

2.1 INTRODUCTION

As a preliminary to any analysis of substitution processes and substitution products it is clearly essential to familiarise oneself with the original product that it is proposed to replace or supplement by a synthetic material. In order to produce a minimum of complications in its use, it will be necessary for a replacement gas to differ as little as possible from the original in a number of characteristics. And in order to assess the most important of these it will be desirable to check which particular performance aspects of original and replacement gas will have to be identical, or at least not very different, to ensure the acceptability of the substitute.

The two gases should behave alike in at least the following respects:

(1) Their physical characteristics should be similar so that they can be stored, piped, liquefied, compressed, expanded, distributed and metered in like manner.
(2) Their chemistry should be similar, *i.e.* the (low) corrosivity, toxicity and (high) stability of natural gases should be matched. Impurities should be removed to the same extent. This will ensure that materials of construction and equipment designed for natural gas can also be used for SNG.
(3) Their performance in combustion equipment should be closely similar in order to guarantee the unchanged operation of burners, appliances, furnaces and other combustion equipment irrespective of a change in supply.

Since the last of these criteria is both complex and of critical importance in regard to the acceptability of a substitute gas, a separate chapter (Chapter 3) will be devoted to the concept of gaseous

13

interchangeability, *i.e.* the definition of combustion parameters that should ensure the replacement of one gas by another without changes in the performance of combustion equipment. In the present chapter it is proposed to discuss and compare only the remaining physical and chemical parameters of natural gases and their components that are significant when one considers the possibility of substitution.

Before doing so, however, we intend to review briefly geological origin, geographical occurrence, composition, transport systems and purification plant for natural gases in various parts of the world. Our purpose is to indicate: firstly the areas that can already be seen as needing either imported or manufactured gas to make up for declining reserves; and secondly to show that SNG cannot be of uniform quality but must take account of differences in gas composition between one producing area and another.

2.2   NATURAL GASES AROUND THE WORLD

A number of comprehensive reviews[1,2,8] have recently been published on the natural gas industry and it is, therefore, not proposed to discuss the sources, the world-wide distribution and the equipment used for exploration and production of natural gas in any great detail.

2.2.1 Sources of Natural Gas[7]

The seepage of natural gas from the soil in certain locations, *e.g.* in China, Mesopotamia and Delphi, was well known to antiquity. However, the eternal flames or the state of intoxication induced by hydrocarbon vapours that resulted were of religious and metaphysical rather than commercial and industrial significance.

Seepage of natural gas or methane, it was found much later, occurred where there were oil deposits or coal measures underground. Fire-damp in coal mines and marsh gas formed from decomposing vegetation were similarly found to consist mainly of methane. There appeared to be some evidence, and this is generally accepted now, that natural gas, just as crude petroleum and coal, is of organic origin and the result of secondary changes that large masses of organic sediments, deposited in geological periods of intensive plant and animal development, underwent in the course of subsequent geological change.*

Another theory, the formation of hydrocarbons by the methanation of carbon dioxide with hydrogen, both of mineral origin, has fewer adherents these days, although it explains fairly satisfactorily the occurrence of different types of deposits such as pure or almost pure methane (dry gas), crude petroleum capped by layers of natural gas, and supercritical fluids consisting of a wide range of hydrocarbons such as methane and light volatile paraffins (gas condensates) or the entire gamut from methane to heavy fuel oil components that must be separated into crude oil and associated gas.

Accumulations of both natural gas and crude oil in geological strata are always characterised by the presence of a layer of impermeable cap rock that prevents the upwards migration of the hydrocarbons. In order to accommodate a significant volume of hydrocarbons the underlying reservoir rock, on the other hand, must be reasonably porous and should be surrounded by or located on top of a permeable hydrocarbon source rock. The accumulation of oil and gas in producible quantities depends on the shape (stratigraphy) of the three layers of different permeability. A dome-shaped configuration is easily recognised by modern exploration techniques and is quite likely to trap migrating hydrocarbons; other promising structures are geological faults, anticlines and other forms of stratigraphic traps.

The geological age of hydrocarbon deposits varies from pliocenic (10 million years) and pleistocenic (1 million years) to some palaeozoic finds (300 million years) with mesozoic rocks as the principal source. However, only sedimentary rocks have been found to be productive, and the occurrence of volcanic activity usually destroys all forms of hydrocarbon deposits.

Commercial production of natural gas in the past had been mainly of associated gas which was co-produced with crude petroleum. However, the search for petroleum, which used the same geophysical criteria as that for gas, inevitably led to some discoveries of dry gas, present as such or associated merely with condensed light hydrocarbons. Finally, some gas finds, especially in Europe, were

---

* In this connection, it is pertinent to mention that methane can be manufactured by anaerobic digestion of agricultural, animal and municipal waste products. Development schemes to probe the commercial possibility of 'syngas' farms and plants are being pursued.[12]

made in fields lying directly above very deep coal deposits, where the absolutely dry nature of the gas confirmed that it had originated from or was at least associated with the underlying coal measures.

2.2.2 Occurrence

Natural gas deposits have by now been found all over the world; but size and importance of these finds varies from the huge gas fields of Texas or Siberia to the relatively small but economically important discoveries in Europe and Japan.[9] A common feature of all discoveries in the proximity of industrial areas and conurbations is their immediate utilisation and rapid exhaustion. Gas fields in remoter areas, much deeper accumulations than those exploited hitherto and discoveries offshore on the Continental shelves of Europe and Australasia have recently been tapped in order to meet the rapidly rising demand.

The constant search for new gas fields and continuing new discoveries make it extremely difficult to provide up-to-date statistics on rates of production and available reserves, particularly since the latter are frequently subdivided into proven, probable and total recoverable volumes, concepts that are both controversial and very difficult to establish, particularly for offshore fields.

Natural gas production and reserves data are therefore generally available only in fragmentary form and the statistics reported in Table 2.1 lay no claim to being either comprehensive or particularly accurate. They do, however, allow one to draw a number of general conclusions.

It is fairly clear, for instance, that some areas such as parts of Western Europe, Japan and the US have not enough natural gas to meet their foreseeable requirements and will have either to limit consumption quite drastically, or to import and synthesise large gas volumes over the next few years. Conversely gas availability in the Middle East, North and West Africa, Indonesia and probably Venezuela by far exceeds local demand. It is also evident that the resources accumulated in the Middle East and African countries for example are of the order of twice those in North America and Canada.

It follows that large gas movements from such areas of surplus availability to other regions of excess demand may possibly develop, particularly since much of the available gas in question is associated with oil production and must therefore be separated if 'stabilised' crude oil is to be moved from the oilfield to a refinery. Separated

TABLE 2.1
Available Natural Gas Statistics—1 January 1974
Reserves, Annual Production and Consumption in 10^9 Nm^3

	Area	Reserves	Production[a]	Consumption[d]
W. Europe	Netherlands	2 606	77·7	31·5
	Great Britain	1 416	28·6	27·5
	Norway	650	—	—
	Germany	348	18·8	30·0
	France	184	10·9	17·9
	Italy	150	13·8	15·7
	Others	82	2·2	15·0
		5 436	152·0	137·6
E. Europe	USSR	20 000	234·0	230·0[e]
	Romania	280	23·2	n.a.
	Poland	140	6·0	n.a.
	Others	170	13·9	n.a.
		20 590	277·1	248·0
North America	USA	7 082[b]	647·0	667·0
	Canada	1 424[c]	92·0	63·5[e]
		8 506	739·0	730·5
Africa	Algeria	3 001	3·4	n.a.
	Nigeria	1 133	15·0	n.a.
	Libya	765	15·9	n.a.
	Others	417	5·9	n.a.
		5 316	40·2	3·1
Latin America	Venezuela	1 189	49·3	n.a.
	Mexico	311	19·4	19·0
	Bolivia	283	4·3	n.a.
	Argentine	226	6·9	n.a.
	Others	573	17·6	n.a.
		2 582	97·5	41·7
Asia and Australia	Australia	1 067	4·2	4·13
	Malaysia and Brunei	566	6·2	6·0[e]
	Indonesia	425	3·6	n.a.
	Pakistan and Bangladesh	506	4·2	n.a.
	Japan	19	2·4	14·0
	Others	640	4·4	n.a.
		3 223	26·0	n.a.

continued overleaf

TABLE 2.1—*continued*

	Area	Reserves	Production[a]	Consumption[d]
Middle East	Iran	7 648	48·0	29·3[e]
	Saudi Arabia	1 441	—	n.a.
	Kuwait and Neutral Zone	1 146	1·4[d]	n.a.
	Iraq	623	4·4	n.a.
	Abu Dhabi	354	—	n.a.
	Others	265	—	n.a.
		11 477	53·8	n.a.

n.a. Not available.
[a]   Total indigenous gas production including commercial gas, re-injected gas and flared gas.
[b]   Including Alaskan reserves.
[c]   Not including 'arctic' gas reserves.
[d]   Only gas sales.
[e]   Includes exports.
Sources: Oil Gas J., March 4, 1974, Dec. 24, Dec. 31, 1973; Iranian Petrol. Inst. Bull., 55; Petrol. Econ., Nov. 1974.

associated gas can, of course, be re-injected and may thereby provide additional driving power for oil production. However, not all oilfields are suitable for re-injection, apart from the massive capital cost of the necessary equipment, and substantial quantities of gas have been and are still being flared at some crude oil production points. It is to reduce the losses incurred by flaring and, of course, in order to meet the gas demand of industrial and domestic consumers that large volumes of gas are already being moved and that even larger gas transport schemes are under consideration.

2.2.3 Transport of Natural Gas

There are basically two ways in which natural gas can be shipped from a gas or oilfield to its consumer. Either it can be moved by compressors along a pipeline or it can be liquefied and the liquid carried by specially designed tankers to its markets.

Pipeline transportation will generally be preferred where shipment is across land rather than sea, where there is little or no likelihood of interference with permanent installations by political or ideological opponents and where there are no serious physical obstacles to pipeline construction. Gas liquefaction, on the other hand, although generally more expensive, is the only possible means of conveyance by ship across deep open seas and may also be preferred for reasons of politics and strategy.

Transport by pipeline requires gas compression; since throughput of a line is a function of pipe diameter and pressure drop—apart from physical properties of the gas—economic optimisation requires careful calculation of pressures, and hence pipe strength, along the line. Compressor stations along the line must be planned in such a way that their extra cost is more than compensated by the additional carrying capacity. Since, inevitably, some pressure drop occurs, careful calculation of so-called retrograde condensation phenomena is required.[3] Condensate formed in the line must be trapped and separated before the gas is re-compressed to prevent damage to compression plant.

Other gas properties that affect behaviour during pipeline transmittal are moisture content, corrosivity caused by impurities in the gas, gas compressibility and viscosity.

If natural gas is to be liquefied and shipped as LNG, dependent on the available pressure, it may have to be re-compressed. Complex liquefaction cycles in which energy consumption is minimised by careful precooling or by a combination of the two refrigeration mechanisms have been designed in recent years, to be used in the massive gas liquefaction facilities that were erected in North Africa, Alaska and Brunei. Many other large liquefaction projects for gas produced in the Persian Gulf, Indonesia, Nigeria and elsewhere are being negotiated.[4]

Gas properties of significance in LNG shipment are again gas make-up, moisture content, impurities and constancy of composition; any change in thermal properties—specific heat, heat of vaporisation, thermal expansion, boiling point or boiling range—will clearly affect the operation of a carefully optimised liquefaction plant. Changes in liquid density have also been shown to result in dangerous conditions of instability, and accidents due to roll-over, the sudden readjustment of layers of different density, can occur if LNG composition undergoes a sudden change.

Major international projects for the transport of natural gas by pipelines and in the form of LNG are listed in Table 2.2 which shows, in conjunction with the data of Table 2.1, that, as one would expect, in most major moves pipeline gas or LNG is shipped from areas of local surplus, such as North Africa, Persian Gulf and Indonesia, to the

TABLE 2.2

Major International Natural Gas Movements

From	To	By
USSR (Urals)	Central Europe	1 420 mm 4 800 km pipeline (under construction)
USSR (Sakhalin)	Japan (Hokkaido)	600 mm 1 500 km pipeline (under negotiation)
Iran	S. USSR	32 in line, 1 000 km pipeline
Holland	Italy	1 line carrying 6×10^9 m^3/year
Holland	Germany	2 lines up to 18×10^9 m^3/year
Holland	Belgium/France	1 main line carrying up to 21×10^9 m^3/year
Norway	EEC	1 34 in line carrying up to 20×10^9 m^3/year
Canada	Central and W. US	Several lines
Algeria	Italy	1 36 in line carrying 11×10^9 m^3/year (planned)
Libya	Italy/Spain	4 LNG vessels carrying $3 \cdot 5 \times 10^9$ m^3/year
Alaska	Japan	2 LNG vessels carrying $1 \cdot 6 \times 10^9$ m^3/year
Brunei	Japan	6 LNG vessels carrying $7 \cdot 8 \times 10^9$ m^3/year
Algeria	Europe I	3 LNG vessels carrying $5 \cdot 0 \times 10^9$ m^3/year
Algeria	US (East Coast), I, II, III	Several LNG vessels for $0 \cdot 5$, 10, $7 \cdot 4 \times 10^9$ m^3/year
Abu Dhabi	Japan	4 LNG vessels carrying $3 \cdot 2 \times 10^9$ m^3/year
Iran	Japan/US (East Coast)	Not fixed, $10 \cdot 6 \times 10^9$ m^3/year
Indonesia	Japan/US (West Coast)	At least 5 LNG vessels carrying $5 \cdot 5 \times 10^9$ m^3/year
Algeria	Europe II	2 LNG vessels carrying $3 \cdot 5 \times 10^9$ m^3/year
Algeria	Spain I, II	Not fixed, $4 \cdot 5 \times 10^9$ m^3/year
Algeria	Europe III	Not fixed, 24×10^9 m^3/year
Malaysia (Sarawak)	Japan	Not fixed, 7×10^9 m^3/year

Additional LNG projects under study envisage shipment from USSR to US, USSR to Japan, Nigeria to US, Australia to Japan, Qatar to Japan.

major deficiency areas, such as the East Coast of the US, Europe and Japan. The rate of development, and the economics of these ambitious transportation schemes are bound to affect the timing of competing gas manufacturing enterprises.

2.3 NATURAL GAS PROCESSING AND PURIFICATION

While some natural gases need little or no clean-up because they consist of almost pure methane or of methane mixed with small proportions of other hydrocarbons, mainly ethane, there are other gases, particularly those associated with liquid petroleum, that require complex processing and purification before they can be compressed for transmission by pipeline or for liquefaction.[1,6]

The most common impurities that have to be separated as soon as possible are liquid hydrocarbons and water. Both complicate handling of the gas, on the one hand by forming hydrocarbon condensates that partly fill the pipeline and result in two-phase flow, an inefficient type of transport, or on the other hand by forming solid hydrocarbon hydrates at high pressure and low temperature, which can block pipelines, valves and other equipment. A third type of impurity, hydrogen sulphide, is corrosive, and has to be removed mainly in order to protect metal parts in contact with the gas.

2.3.1 Dehydration

The removal of liquid water and water vapour from natural gas can be effected by refrigeration, by liquid absorption in a non-volatile solvent, by adsorption on a solid desiccant or by a combination of two or more of these routes. Refrigeration removes both water and heavier hydrocarbons and it is common practice to remove both types of impurity simultaneously.

The most common solvent employed to dehydrate natural gas under pressure is triethylene glycol. This boils at 287·4 °C at atmospheric pressure and when used in the process its water content varies from 3 to 5 % wt 'dry' to over 10 % wt 'wet'. Wet glycol is dried by fractionation at atmospheric pressure, water vapour leaving as overhead and dry glycol as bottoms. The degree of drying depends on the circulation rate of glycol solution; assuming that 97 % wt solution enters the scrubber and glycol concentration at the exit is 90 % wt, then a flow of 12·9 kg of absorbent will be needed to remove 1 kg of water vapour from the gas.

Activated alumina reduces the moisture content of natural gas even more effectively than does triethylene glycol. It therefore is used widely especially in large natural gas purification plants. The adsorption process is operated at high pressure, sometimes with external cooling to remove the heat evolved. The water content of the saturated adsorbent is 9–11 % wt and is removed by a reverse flow of gas preheated to about 300 °C, through the adsorbent bed. Other desiccants such as molecular sieves or zeolites can be used for the removal of water, simultaneously with hydrocarbons and sour gas components, depending on the type of sieve and operating conditions.[10] The regeneration conditions generally however are more severe than those required for alumina.

Dehydration by refrigeration, in the presence of a hydrate suppressant, can also remove from the gas both water vapour and higher hydrocarbons, and by more intensive temperature reduction one can also remove CO_2 and sulphur compounds. Just as in LNG production, either a Joule–Thompson expansion through an orifice or an expansion engine producing external work can be used. The plant differs from an LNG plant, however, in the heat exchange between incoming and exit gas which precools the former after compression and heats the latter before dispatch.

Condensed water and hydrocarbons are separated from the cooled gas and allowed to settle so that water can be decanted and hydrocarbon condensate recovered.

2.3.2 Removal of Heavier Hydrocarbons

The removal of heavier hydrocarbons and 'permanent' gases is usually carried out in a number of steps. Crude oil produced under pressure is normally saturated with water and gas and therefore has to be first decanted, to remove liquid water, and then 'stabilised' by stepwise expansion to remove first high-pressure, then medium-pressure, then low-pressure gas. The three gas streams, which contain varying amounts of heavier hydrocarbons, are, as a rule, combined and subjected to various forms of absorption or de-gasolining treatment.

The standard means of separating what is variously called gas condensate, casing head or natural gasoline, is to scrub the gas with a refrigerated absorbent, usually a hydrocarbon oil. After dehydration in a glycol tower the gas is expanded and cooled to about −5 °C to remove residual moisture and glycol. It is then passed counter-currently through a scrubber through which absorbent oil, cooled to the same low temperature, descends and removes from the gas most of its high-boiling constituents.

The latter have to be recovered by heating the enriched absorbent oil to 135–140 °C and fractionating it into a heavy lean component and a gasoline overhead, which itself can be separated into propane, butanes and residual gasoline. Recycle lean oil is cooled by heat exchange with dry exit gas and other streams before it is re-used in the gas scrubber.

Under these circumstances the de-gasolined natural gas consists essentially of methane, smaller amounts of ethane, any traces of permanent gases such as nitrogen, oxygen, helium which happened to be present in the crude gas, and residual acidic impurities such as sulphur compounds and carbon dioxide.

2.3.3 Desulphurisation and CO_2 Removal

The removal of acidic gases, also called sweetening, becomes essential if hydrogen sulphide content exceeds values specified in pipeline contracts, which are often as low as $0.0055 \, g/Nm^3$. However, only if H_2S content is higher than about $5g/Nm^3$ does it become attractive to recover elemental sulphur from the extracted H_2S. Carbon dioxide removal, on the other hand, is only essential if it is present in massive concentrations, i.e. when the calorific value of the gas is seriously reduced or when there is a risk of line corrosion by wet pipeline gas. Reduction of CO_2 to about 3% vol appears to be normal under these circumstances. Complete removal of both H_2S and CO_2 is, however, required if the natural gas is to be liquefied, since solidified material, the freezing points of CO_2 and H_2S being higher than those of the other gas components, would interfere with the normal working of the plant.

The standard process for the removal of H_2S from natural gases uses an aqueous solution of monoethanolamine. The gas to be treated is passed counter-current to the solution through an absorption tower and leaves the process plant free of H_2S after heat exchange with the incoming lean absorbent. The rich solution leaves the bottom of the absorption tower, and after heat exchange with various streams enters a stripping tower. In the latter, the injection of live steam and a bottoms recycle through a boiler remove hydrogen sulphide from the amine solution. Acid gases are cooled, water vapour carried over with the gases is condensed and the residual gases, depending on volume and sulphur concentration, are either flared or processed for sulphur recovery.

The removal of CO_2, while possible by means of amine solution, is more economically carried out, at least in those instances where CO_2 concentration is 10% vol or higher, by scrubbing the gas with hot potassium carbonate solution (Benfield process) or by a similar extraction route such as Vetrocoke or Catacarb, each of which uses different additives to improve hydration rate of the gas in the potassium carbonate solution.

The spent absorbent is again withdrawn from the bottom of the scrubber and regenerated at atmospheric pressure by steam stripping. Carbon dioxide gas, after partial water separation, is vented to atmosphere, and the regenerated potassium carbonate solution is recycled to the absorber tower.

The process of sweetening may also be carried out on the dried gas by a molecular sieve adsorption process which can be so designed as to leave small volumes of carbon dioxide in the product gas.[10]

2.3.4 Removal of Nitrogen

It is rarely necessary to separate nitrogen from natural gas. As an inert diluent, nitrogen admittedly takes up pipeline volume and consumes compression energy, but the cost of nitrogen removal is usually higher than the potential gain, and except in one instance, the Alfortville separation plant near Paris, nitrogen removal from natural gas has been avoided. Even at Alfortville, however, separation economics would have been doubtful had it not been for the simultaneous recovery of helium.

The plant in question is complex. Pipeline gas from Holland, containing 14% wt nitrogen, first enters a CO_2 removal section; it is then carefully dehydrated by means of triethylene glycol and high-boiling components are removed by liquid ammonia chilling and primary fractionation in a low temperature, medium pressure separator. The processed gas at that stage consists of methane, ethane, nitrogen, and helium, which are subsequently split into three streams by low-temperature fractionation. It is claimed that once the unit is operating, heat exchange between incoming and outgoing streams is sufficient, with the aid of a minimum of Joule–Thompson refrigeration, to provide the energy requirements of the plant.

2.4 THE CHEMICAL AND PHYSICAL PROPERTIES OF NATURAL GASES

There is no doubt that one could, in theory, by means of extremes of processing and purification, reduce practically all naturally occurring gaseous fuels to pure methane. However, little would be served by such standardisation since different types of natural gas will only be used interchangeably on rare occasions. In practice fairly wide variations in properties and composition will occur, and it is proposed in the following section to discuss the various gas characteristics and their range of values.

Since many properties of gaseous mixtures are simple averages of the characteristics of their components, little need be said apart from stating the composition of the gas together with its density, molecular weight, calorific value, flame temperature, heat of vaporisation, and compressibility all of which are approximately weighted averages of the corresponding parameters of the individual gas components. Other characteristics of gas mixtures such as Wobbe Index, inflammability range, burning velocity, boiling point, critical point cannot be evaluated as simple weighted averages but require a more complex approach. However, it is generally true that in order to assess the characteristics of a natural gas it is essential to know firstly its composition and secondly the relevant properties of the individual components.

It is not proposed to list comprehensively all the physical and chemical characteristics of natural gases and their components but to confine our attention to those parameters that have a bearing on the replacement or supplementation of a gas. Apart from listing the properties concerned we shall therefore indicate their significance in regard to the manufacture and required characteristics of SNG.

2.4.1 Natural Gas Characteristics[5,6,11]

Specific gravity of a gas is usually expressed with reference to air, the latter being assigned the value of unity. Specific gravity affects the design and energy consumption of compressors, flow rates through pipes, valves and orifices, buoyancy and therefore shape of flames, design of gas holders and most other gas-handling equipment. (Very recently, the term 'relative density' has come to be preferred whereby the density of a gas at a named temperature and one atmosphere pressure is referred to that of air, being unity.)

The *specific gravity* of a liquid, usually expressed with reference to water, is of significance when natural gases are liquefied. Design of tanks, pipes, valves, etc. is affected. It is not a linear function of composition.

Molecular weight is a basic parameter. At atmospheric pressure and ambient or higher temperature it is linearly related to gas density.

Specific heats at constant volume and constant pressure affect temperature changes during compression, expansion, cooling or heating and therefore the design of compressors, expanders and heat exchangers.

Compressibility data are required in connection with gas compression and the design of compressors and pressure vessels. The compressibility factor, *i.e.* the divergence from ideal gas behaviour, allows one to calculate gas density at pressures higher than atmospheric. Its value varies with temperature.

Viscosity of a gas affects the pressure drop incurred when it passes through a pipe. It increases with increase in temperature and thus increases the resistance to flow of a gas when it is heated. Viscosity of a liquid, on the other hand, decreases with increase in temperature, and is significant where LNG is pumped.

Latent heat of vaporisation is only important where natural gases are liquefied or LNG is vaporised.

Boiling point of a liquid is only clearly defined for individual components, since liquid mixtures boil—and gaseous mixtures condense—over a range of temperatures. Since boiling point is the temperature at which vapour pressure and external pressure are in equilibrium it is clearly pressure-dependent. Furthermore, since liquids do not mix ideally, boiling point is not a linear function of composition.* Boiling points are again significant in gas liquefaction.

Calorific value of a gas is the heat generated by burning unit volume of the gas with air under standard conditions, involving either condensation of water vapour formed during combustion (upper or gross calorific value) or leaving the water in the vapour state (lower or net calorific value).

Gas–air ignition limit is the range of concentrations over which gas/air mixtures can be ignited by means of a spark. The range,

---

* Numerous rules have been developed for the calculation of vapour pressures and boiling points of non-ideal solutions. Mixtures of hydrocarbons conform fairly accurately with the Benedict–Webb–Ruben rules. The presence of non-hydrocarbons in substantial concentrations necessitates more complex correlations.

particularly the upper or rich limit, depends on temperature and pressure. Its significance lies mainly in the design of burners and appliances with stable flames, but it also has a bearing on safety; gas leakage must not result in gas/air mixtures falling within the ignitable range. Neither of the ignition limits is a linear function of gas composition.

In mixtures composed entirely of inflammable components the limits of inflammability:

$$L = \frac{100}{\dfrac{a}{A} + \dfrac{b}{B} + \dfrac{c}{C} + \cdots}$$

where $a, b, c, \ldots$ are the volume percentages (on an air-free basis) and $A, B, C, \ldots$ are the inflammability limits (lean and rich respectively) of the pure components. The complex effect of inerts must be allowed for in the evaluation of $A, B, C \ldots$.

Flame temperature (F) of a gas/air mixture will be at a maximum if fuel and oxidant are in stoichiometric proportions. It is a measure of the performance of a gaseous fuel in high temperature applications such as welding and cutting but also has a bearing on heat transfer efficiency in ordinary combustion equipment. Provided the effect of inerts has been duly taken into consideration the flame temperature of a mixture approximately equals that of the weighted average of the components.

Some of the properties mentioned have been listed for various natural gases[11] and are shown in Table 2.3.

2.4.2 Composition of Natural Gases

The main component of all natural gases is methane and a large number of piped gas systems are based on gases containing very low concentrations of gases other than methane. However, there are exceptions, and particularly if we consider natural gases as they emerge from the well, *i.e.* before processing and purification, components other than methane are often present in significant proportions.

A further difference in composition may arise as a result of liquefaction of a gas. Since higher hydrocarbons condense at higher temperature than does methane, their concentration in LNG may be higher than in the uncondensed gas. Similarly nitrogen, owing to its lower boiling point than that of methane, may be reduced in concentration as a result of liquefaction.

TABLE 2.3
Some Properties of Various Natural Gases

	Groningen	Lacq	Libyan[a]	North Sea[b]
Molecular weight	18·62	16·43	22·87	16·96–18·95
Relative density, 15 °C, 1 atm (air = 1)	0·643 8	0·568 2	0·7927	0·586 5–0·655 3
Density, kg/m^3, 15 °C, 1 atm	0·789 0	0·696 4	0·9714	0·711 8–0·803 1
Calorific value, 15 °C dry				
gross, Btu/ft^3	891	1029	1 369	1 019–1 052
net, Btu/ft^3	803	927	1 245	921–949
Wobbe Number dry Btu/ft^3	1 110	1 363	1 538	1 325–1 359
Weaver flame speed factor	12·93	14·12	15·04	13·75–14·11
Inflammability limits, % in air				
lower	5·7	4·9	3·8	4·9
upper	16·8	15·0	13·5	15·6

[a] LNG.
[b] Includes West Sole, Viking, Leman Bank, Hewett and Indefatigable fields.
Courtesy: British Gas Corporation.

The liquefaction process, incidentally, can be carried out stepwise and this provides an opportunity for the removal of higher hydrocarbons; alternatively the complete gas mixture can be liquefied and the resultant LNG will then contain relatively high concentrations of ethane, propane and butane.

Certain natural gases contain substantial concentrations of impurities such as carbon dioxide or hydrogen sulphide which are normally removed or at least reduced before transmission by pipeline.

The gas compositions listed in Table 2.4 are thus not all typical of pipeline gases but at least in part represent extremes of composition that, as far as most gas distributors are concerned, are mainly of academic interest. On the other hand the table shows that even within the range of acceptable pipeline gases considerable variations in composition can occur.

2.4.3 Physical Properties of Natural Gas Components

Since almost all the characteristics of gas mixtures can be predicted from composition data, and from the corresponding parameters of the components—either by simple linear interpolation or by some more complex mathematical method—it is desirable to have accurate and comprehensive physico-chemical data available for all these components. Such data have in fact been determined very accurately over the past few years and Table 2.5 is based on recently published values of those parameters that are considered important in the context of SNG manufacture. They refer to all the principal components of natural gases that are present in sufficient concentrations to affect the properties of the mixed gas.

Small concentrations of helium, argon, benzene, etc., while they can easily be detected by the usual chromatographic analysis methods, will have normally no practical effect on the characteristics of the blend.

Further details of the composition and properties of European natural gases, and of the physical characteristics of the pure components can be found in a recently published Data Book.[11]

2.5   CONCLUSIONS

While all natural gases consist mainly of methane there are present in addition a number of other components which owing to their range of

TABLE 2.4
Composition of Natural Gases (% vol)

Components	Rundle, Crossfield	Amarillo, Texas	Monroe, Louisiana	Lacq, Pyrenees (c)	Lacq, Pyrenees (d)	Groningen, Holland	Cortemaggiore, A level	Ravenna, Offshore
Methane	80·60	72·9	94·7	69·1	97·4	81·3	93·28	99·52
Ethane	6·45	19·0	2·8	2·8	2·2	2·85	4·14	0·06
Propane	2·30	—	—	0·8	0·1	0·39	1·29	0·01
Butanes	1·23	—	—	1·5	0·05	0·13	0·58	trace
C_5 and higher	1·27	—	—	0·6	—	0·06	0·41	—
Carbon dioxide	6·13	0·4	0·2	9·7	—	1·0	trace	0·01
Hydrogen sulphide	1·06	—	—	15·0	—	—	—	—
Nitrogen	—	7·7	2·3	0·5	0·3	14·3	0·30	0·40
Helium	0·96	—	—	—	—	0·05	—	—
TYPE	USA, sour associated high He	USA, sweet dry high N_2	USA, sweet dry	France, dry		Dutch, sweet dry high N_2	Italy, sweet dry	Italy, sweet dry almost pure CH_4

TABLE 2.4—continued
Composition of Natural Gases (% vol)

Components	North Sea[e]	Hassi R'mel, Algeria	Arzew,[b] Algeria	Brega,[a] Libya	Kapuni, New Zealand	La Paz, Venezuela	Flare gas, Kuwait	Bass Straits, NSW
Methane	93·63	76·66	86·3	66·8	44·2	78·1	72·2	93·2
Ethane	3·25	9·36	7·8	19·4	6·1	9·9	15·5	1·9
Propane	0·69	2·84	3·2	9·1	3·4	5·5	6·8	1·1
Butanes	0·27	1·77	0·6	3·7	1·6	2·8	2·5	—
C_5 and higher	0·20	4·28	0·1	<1·0	0·4	2·1	3·0	0·9
Carbon dioxide	0·13	0·20	—	—	44·2	0·4	—	—
Hydrogen sulphide	—	—	—	—	—	—	some	—
Nitrogen	1·78	4·96	2·0	—	0·1	1·2	some	2·9
Helium	0·05	—	—	—	—	—	—	—
TYPE	UK, sweet dry	N. Af., sweet dry	N. African LNG (ex LPG)	N. African LNG (total)	dry high CO_2	sweet associated	M.E., sour associated	Australian sweet, dry
	—	—				—		—

[a] Composition of LNG ex Brega liquefaction plant.
[b] Composition of LNG ex Arzew liquefaction plant.
[c] Crude.
[d] Purified.
[e] Average received, Bacton terminal see ref. 11.

TABLE 2.5
Selected Physical Properties of Natural Gas Components

	Methane	Ethane	Propane	Butanes		Pentanes		Nitrogen	CO_2	H_2S
				Iso	Normal	Iso	Normal			
Molecular weight	16·04	30·07	44·10	58·12	58·12	72·15	72·15	28·02	44·01	34·08
Freezing point (1 atm abs) °C	−182·5	−183·3	−187·7	−159·6	−138·3	−159·9	−129·7	−210·0	−78·50[b]	−85·60
°F	−296·5	−297·9	−305·8	−255·5	−217·0	−255·8	−201·5	−210·0		−122·10
Boiling point (1 atm abs) °C	−161·6	−88·6	−42·1	−11·7	−0·5	27·8	36·1	−195·8		−60·30
°F	−258·9	−127·5	−43·7	10·9	31·1	82·4	96·9	−320·4		−76·50
Density of liquid (at vapour pressure) S.G. 15/15 °C	0·3	0·3771	0·5077	0·5631	0·5844	0·6248	0·6312		0·914[c]	
Density of gas (5 °C, 1 atm abs) Air = 1[a]	0·554	1·038	1·522	2·066	2·066	2·491	2·491	0·967	1·5494	1·1763
lb/1 000 ft³[a]	42·27	79·23	116·19	153·15	153·15	190·11	190·11	73·90	116·70	90·07
g/m³	716·8	1 356·0	2 019·0	2 668·0	2 703·0	3 195·5	3 216·5	1 250·6	1 976·3	1 521
Critical temperature °C	−82·6	32·4	96·8	135·0	152·0	187·2	196·4	−147·0	31·1	100·0
°F	−115·8	90·3	206·3	275·0	305·6	369·0	385·5	−232·8	88·0	212·0
Critical pressure atm	45·8	48·3	42·0	36·5	37·9	32·9	33·3	33·6	73·8	89·4
psi[a]	673·3	710·0	617·4	536·6	557·1	483·6	490·0	493·9	1 084·9	1 314·2
Calorific value (at 15 C, gross) Btu/scf[a]	1 010	1 769	2 517	3 253	3 262	4 000	4 009			673·7
kcal/Nm³[a]	9 643	16 390	24 032	31 060	31 145	38 192	38 278			
Btu/lb[a]	23 885	22 323	21 664	21 238	21 299	21 041	21 088			
kcal/kg	13 271	12 403	12 037	11 800	11 834	11 690	11 716			
(at 15 C, net) Btu/scf	895	1 580	2 280	2 960	2 967	3 680	3 680			580
kcal/Nm³	8 551	15 095	21 783	28 280	28 347	35 159	35 159			5 541
Inflammability limit (% in air) lower	5·0	2·9	2·1	1·8	1·8	1·4	1·4			
upper	15·0	13·0	9·5	8·4	8·3	8·3	8·3			
Heat of vaporisation (at atm boiling point) Btu/lb	219·2	210·4	183·1	157·5	165·7	147·1	153·6	86	101·03[d]	
kcal/kg	121·8	116·9	101·7	87·5	92·1	81·7	85·4	47·8	56·75	
Specific heat (at 15, 1 atm) Cp gas[a]	0·5271	0·4027	0·3885	0·3872	0·3908	0·3827	0·3883	0·248 2	0·1801	
Cv gas[a]	0·403	0·344	0·343	0·353	0·357	0·355	0·361	0·177	0·153	
(at atm boiling point) Cp liquid	0·925	0·926	0·592	0·570	0·564	0·535	0·542			

[a] Assumes ideal gas behaviour.
[b] Undergoes sublimation.
[c] g/ml at 0 °C and 34·3 atm.
[d] At 0 °C.

concentrations and variety, can produce gas mixtures of widely varying characteristics. In fact a review of the world's gas reserves and production shows that there are some large gas deposits of unusual composition around the world. Even if these gases are processed and purified there often remain substantial differences in gas characteristics.

To make a synthetic material genuinely equivalent to the natural product one has to ensure that all important properties of the original and its replacement are sufficiently alike to guarantee the same performance in use. Characteristics of specific natural gases that are to be replaced by synthetics will therefore have to be examined very carefully and those parameters that affect interchangeability will have to be noted.

However, since the characteristics of gaseous mixtures are to a large extent linear averages of the corresponding parameters of the individual components, it is of even greater importance to establish accurate information on the one hand on gas composition, on the other on the physico-chemical properties of all the constituents and also to establish calculating methods in order to predict those characteristics of mixtures that are not simple weighted averages of those of the individual components.

The gas parameters dealt with in the present chapter specifically exclude the so-called combustion characteristics, which define combustion behaviour in certain burners and appliances. Owing to their complexity and their importance in regard to substitution of natural gases by replacement materials, it is proposed to review combustion interchangeability separately and to devote the next chapter to that topic.

REFERENCES

1. Medici, M. (1974). *The Natural Gas Industry*, Newnes–Butterworth, London.
2. B.P. Trading Ltd (1972). *Gas Making and Natural Gas*, B.P. Trading Ltd, London.
3. Dick, M. N., Gandhi, J. R. and Lom, W. L. (1973). The prediction of retrograde condensation in natural gases, *Gas Wärme Int.*, **22**(2), 70–78.
4. Lom, W. L. (1974). *Liquefied Natural Gas*, Applied Science Publishers, Barking.

5. American Gas Association (1969). *Gas Engineer's Handbook*, Industrial Press Inc., New York.
6. Katz, D. L. *et al.* (1959). *Handbook of Natural Gas Engineering*, McGraw-Hill, New York.
7. Laurie, J. (1961). *Natural Gas and Methane Sources*, Chapman and Hall, London.
8. Hepple, P. (1973). *Outlook for Natural Gas—a Quality Fuel*, Applied Science Publishers, Barking.
9. Tiratsoo, E. N. (1972). *Natural Gas*, 2nd Ed., Scientific Press Ltd, Beaconsfield.
10. Schoops, R. J. (1967). Molecular sieves treat natural gas, *Oil Gas Internat.*, **7**(7), 30.
11. British Gas Corporation (1974). *British Gas Data Book*, Vol. 1.
12. Institute of Gas Technology (1975). *Gas Scope*, No. 31, Chicago.

Chapter 3

The Interchangeability of Fuel Gases

3.1 INTRODUCTION

The principal reasons for the preference shown for gaseous fuels, compared with liquid or solid fuels, by industrial, commercial and particularly by domestic consumers are the following:

the cleanliness of gaseous fuels, *i.e.* the absence of sulphur oxides and other impurities in the combustion gases,
their ease of combustion, *i.e.* the absence of partly oxidised intermediates such as carbon monoxide, formaldehyde etc... in the combustion gases, even in the absence of excess air,
their ease of control, *i.e.* simple regulation of flow of gas and oxidant,
their stability, *i.e.* the absence of soot or carbon deposits on mixers, burners and heated surfaces.

In order to meet the high standards indicated in the above list it is essential that all the relevant characteristics of a gaseous fuel supply should be absolutely uniform. Specifically, a change from one gas to another, *e.g.* from natural gas to SNG, must not result in any changes in the performance of combustion equipment, in the frequency of burner maintenance or in the adjustment of automatic controls. This is in fact what is implied by the stipulation that the original gas and any acceptable substitute must be fully interchangeable.

Clearly, if it is proposed to change from one gas to another, it will not be possible to examine individual burners and to ensure that the replacement gas performs equally well as the original in each of them. In order to ensure a reasonable degree of interchangeability without extensive testing, one will therefore have to define a number of fuel parameters, the value of which must not vary over more than a limited range. In the interest of lowest cost and widest possible acceptability of substitute gases, the number of these parameters must be low and the permissible tolerances as wide as possible. But, clearly, a balance will have to be struck between producing an expensive substitute that would be identical with the original and, therefore, suitable for every single unit in a large burner population and the lower cost of producing an SNG, whose performance is acceptable, though perhaps not identical, in all combustion equipment.

The argument as to what constitutes acceptable interchangeability has not so far been fully resolved. A number of systems and standards have been set up in different countries, partly as a result of the preponderance of different makes of combustion equipment, partly because basically different gases—coal gas, natural gas, cracked oil gas—had set the original standards. However, most of the interchangeability parameters used by the gas distribution industry refer to a limited number of performance characteristics, and in general one can state that gases are considered interchangeable provided they are equivalent in the following respects:

at constant supply pressure the rate of flow of heat energy through an orifice or burner remains reasonably constant,
at constant supply pressure and flow rate of gas and oxidant the size and shape of the resultant flame are reasonably steady,
at constant air/fuel ratio and flow rate the formation of partially oxidised intermediates does not exceed a certain maximum,
at constant air/fuel ratio and flow rate the formation of soot and carbon does not exceed a certain maximum.

The significance of these four criteria is not the same; the most important is the first, and practically all interchangeability systems include some means of measuring the flow of heat energy, a topic that will be discussed in greater detail later on. The second, which defines size and shape of a premixed gas flame, is a function of the flame speed; the precise value of the latter being similar for different paraffinic hydrocarbon gases, methane, ethane, etc., it is of importance mainly as far as differences between hydrocarbons and hydrogen-containing gases are concerned. Finally, the formation of combustion intermediates and soot can be significant where fuel gases contain unsaturated intermediates, and soot can be significant where fuel gases contain unsaturated and higher boiling hydrocarbons or aromatics; in all other instances carbon deposits and pollutant formation will not exceed values considered acceptable with natural gases and established combustion equipment. In consequence, insistence on this aspect of interchangeability is generally confined to distribution areas where synthetic or coal-derived gaseous fuels have been used in the past.

In this chapter it is therefore proposed to expand first on the concept of thermal energy flow through a burner or orifice and to discuss the concept of a Wobbe Index and a number of derived or similar functions that have been developed to express thermal output. It is next intended to deal with burning velocities of gases and their effect on flame size and shape. Next we will discuss the remaining interchangeability parameters concerned with soot and combustion intermediate formation. Finally we plan to compare a number of different interchangeability systems; that is sets of gas characteristics that have been used at different times and in different countries to express gaseous interchangeability.

3.2 THE FLOW OF HEAT ENERGY—THE WOBBE INDEX

The combustion of unit volume of a fuel gas under standard conditions of pressure, temperature and humidity, releases a quantity of heat energy known as the calorific value of the gas. If water vapour formed in the process is condensed the heat released equals the gross or upper calorific value; if the water vapour is left in the vapour phase the heat released equals the net or lower calorific value. If the fuel in question is sold by volume it is consequently essential for equitable pricing that the calorific value, more importantly the net than the gross, should be kept constant at all times, irrespective of changes in supply or source.

On the other hand, if gas supplies are invoiced in terms of heating value, this is no longer essential; identical calorific value is therefore not included in the concept of technical interchangeability, but it is often desirable in order to ensure the commercial interchangeability of two or more gases. It may, for example, be desirable to supply a gas of higher calorific value in order to meet other interchangeability criteria; if there is no provision for increasing the volumetric price under these circumstances, this means that value is given away by the supplier.

Constant calorific value, on the other hand, does not by itself ensure that a constant quantity of energy is released in a burner. The flow of gas through a pipe, an orifice or a valve of fixed dimensions is a function of pressure difference, viscosity (of less importance) and specific gravity. If the first two are maintained constant, rate of flow becomes inversely proportional to the square root of the specific gravity. For a constant supply of heat energy through a given burner, assumed to be equipped with a pressure regulator, it will be essential that at all times the function 'calorific value$/\sqrt{\text{specific gravity}}$', normally designated as the 'Wobbe Index' (WI), should be constant.

Clearly if supply pressure changes as well, then, since rate of flow is approximately proportional to the square root of the pressure drop, the function 'calorific value $\sqrt{\text{excess pressure}}/\sqrt{\text{specific gravity}}$' termed the 'extended wobbe index' should remain unchanged. In other words:

$$\text{WI extended} = \text{WI} \times \sqrt{P} = \text{Constant.}$$

where P = excess pressure.

The WI as used in combustion calculations is normally based on the upper or gross calorific value. However, since water vapour in the combustion gases is not normally condensed, it would be more logical to think in terms of net calorific value and to calculate a net WI from it.

Finally, to allow for certain irregularities in the combustion performance of higher hydrocarbons, on the one hand, and of gas mixtures containing oxygen or CO_2, on the other, semi-empirical corrections to both Wobbe Index and Extended Wobbe Index have been introduced to derive a so-called 'Modified Wobbe Index'.* Use of the latter term is said to permit a more accurate prediction of faulty combustion in French appliances.

The three concepts: Wobbe Index, Extended WI, and Modified WI are thus measures of thermal throughput and therefore of the capacity of any burner or combustion system. A change of gas supply

---

* WI modified = WI $\times k_1 \times k_2$

$k_1 = 1 - \dfrac{12 \cdot 5 (CO + 4O_2 - 0 \cdot 5 CO_2)}{CV}$

$k_2 = 1 + 5 \cdot 8 \times 10^{-6} CV^* + 0 \cdot 131 \times 10^{-9} CV^{*2} - 1 \cdot 22 \times 10^{-15} CV^{*3} + \ldots$

where CV^* is the gross calorific value of hydrocarbons higher than CH_4.

that results in a lowering of the Wobbe Index will accordingly reduce system capacity and may be unacceptable for that reason.

Undoubtedly the most important and most widely used combustion system is the low pressure, partial premix, atmospheric or, for short, Bunsen burner. In a Bunsen burner, gas is released through an orifice into an open-ended mixing tube. Air is entrained owing to the gas pressure reduction after the orifice, and mixing of the gas with primary air takes place in the tube, which can be either parallel sided or a venturi tube. The gas/air mixture burns at the outlet of the mixing tube where secondary air is available.

Between 35 and 60% of the stoicheiometric air requirement is entrained in the tube since this range of air/gas ratios produces a well mixed, stable and relatively noiseless flame and ensures complete combustion without excessive carbon monoxide formation. The volume of entrained air can be regulated by means of a shutter, but it is clearly desirable to minimise the need for shutter adjustment by minimising changes in primary air demand.

The degree of primary aeration of a flame is inversely proportional to the calorific value of the fuel gas, since, as shown in Table 3.1, stoicheiometric air requirement is directly proportional to calorific value, and primary aeration is its reciprocal. Primary aeration is also directly proportional to the momentum of the incoming gas which itself is proportional to the square root of gas pressure and to the square root of specific gravity:

$$\text{Primary aeration} = k_1 \frac{\text{momentum of gas stream}}{\text{calorific value}}$$

$$= k_2 \frac{\sqrt{\text{pressure drop}} \times \sqrt{\text{specific gravity}}}{\text{calorific value}}$$

$$= k_2 \frac{\sqrt{P}}{\text{WI}} \quad \text{where } P = \text{pressure drop}$$

It follows that Wobbe Index, and Modified Wobbe Index, can also be used as a measure of the degree of primary aeration of a gas in a Bunsen burner since the latter is inversely proportional to the former. In other words, maintaining the Wobbe Index of a gas supply constant will not only ensure constant thermal throughput but also constant primary aeration, provided, of course, supply pressure remains unchanged.

The effect of increasing the Wobbe Index of a supply will thus not only be an increase in thermal throughput but also a lowering of

TABLE 3.1
Calorific Value and Stoicheiometric Air Volume

Gas	Gross calorific value, Btu/scf[a]	Stoicheiometric air reqmt., vol/vol	Calorific value/ stoich. air	
Hydrogen	343·2	2·38	144·20	
Carbon monoxide	341·1	2·38	143·36	
Ethylene	1673·3	14·28	117.18	
Methane	1067·2	9·52	112·10 ⎫	
Ethane	1855·6	16·66	111·38 ⎬	average
Propane	2660·0	23·80	111·76 ⎭	111·67
Butanes	3448·4	30·94	111·45	

[a] Gross at 0 °C and 760 mm Hg for dry gas.

primary aeration. A decline of the latter below a critical value, which varies from burner to burner and between appliances, will result in incomplete combustion and the presence of toxic materials, particularly carbon monoxide, in the combustion products. When replacing an existing gas supply by SNG it is thus essential from a safety standpoint to ensure that the replacement gas is not too high in Wobbe Index.

If one wishes to maintain thermal throughput, any increase in Wobbe Index, it will be remembered, must be matched by a corresponding drop in gas pressure. However, to maintain primary aeration one has to increase the square root of the supply pressure proportionately.

It is, thus, not possible to compensate for a change in Wobbe Index by simply altering the supply pressure.

In the light of the importance of Bunsen type burners in any gas burner population it will, therefore, be obvious that constancy of the Wobbe Index of a gas supply must be considered essential and that every effort must be made to ensure that SNG is manufactured to the Wobbe standard set by the original supply.

3.3   THE BURNING VELOCITY OF GASES

When a gaseous fuel emerges from an injector jet, mixes with an oxidant, and burns in the shape of an open flame, a series of very complex events takes place more or less simultaneously. Fuel and oxidant molecules diffuse into each other, heat is transferred from the combustion zone to the unburnt gas, chemical energy is released and high energy molecules, active centres or chemical intermediates—depending on the particular combustion theory that is used—transfer excess energy from reacted molecules to the fresh reactants.

While a detailed analysis of flames is clearly beyond the scope of this book a number of simplifying assumptions permit one to appreciate and make use of the concept of burning velocity or flame speeds. One can claim, for example, that a steady free flame burning in the shape of a well defined surface is simply the result of the regular flow of reactants towards the flame zone balanced by an equal and opposite burning velocity. One can furthermore assume that the only significant—as far as flame stability is concerned—direction of burning is at right angles to the flame front, and for an atmospheric premixed burner, a steady flame implies that the flame velocity component along the axis of the burner tube (S_L) is equal and opposite to the velocity of flow of the gas mixture (V).

The numerical value of S_L also depends on the angle (θ) between the flame front and the burner axis (see Fig. 3.1).

$$S_L = \frac{S}{\sin \theta} = V$$

It follows that if V remains constant for two gases of different flame speeds $(S_1,$ high; $S_2,$ low), θ_1 must be greater than θ_2. This means that a gas with a high burning velocity will result in a flat flame (θ_1 large), and a gas with a low burning velocity will give rise to a flame with a tall thin cone (θ_1 small).

By measuring the value of θ, the flame speed can be determined, provided also that the flow velocity (or volume flowrate and burner tube diameter) is known.

S (flame speed) $= V$ (flow velocity) $\times \sin \theta$ (angle between flame front and burner axis).

It also follows that if there is an incompatibility between V and S, one of two things may happen. If V is too low to match S_L, even if the flame is very flat, the flame front will eventually collapse inside the Bunsen tube, the flame will strike or flash back and establish itself on the injection jet. If S_L, on the other hand, is too low even for a very long flame, the flame front will detach itself from the burner rim and eventually the flame will blow off and be extinguished. In between these two extremes, however, a balance between longitudinal flame

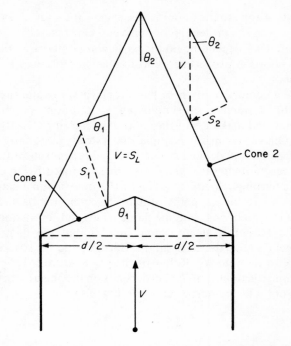

FIG. 3.1   Effect of gas burning velocity on cone height in a Bunsen burner. V, velocity of mixture flow; S_1, flame speed (high); S_2, flame speed (low): S_L, axial component of flame speed; d, burner tube diameter.

speed and velocity of gas flow can be established and a stable flame front will result.

Since flame speed or burning velocity is quite characteristic for a given gas (see Table 3.2) it follows that burner and appliance design must ensure that at normal operating conditions gas velocity and flame speed are matched and that there is scope for adjustment in either direction without getting into a regime of flash-back or blow-off. If an established gas supply is replaced by SNG it is therefore essential that the replacement gas should have a burning velocity not too different from that of the original. Fortunately, as shown in the table, differences in flame speed between saturated hydrocarbon gases are relatively minor and most substitute gases, unless they contain high proportions of hydrogen, will meet this requirement.

TABLE 3.2
Correlation between Various Flame Speed Parameters

Gas	Max. burning velocity in air (cm/sec)	Weaver flame speed factor[a]	Delbourg combustion potential[b]	Holmqvist characterisation factor[c]
Hydrogen	225	100	379	1·102
Methane	34	14·0	40·3	9·73
Ethane	42	16·0	73·3	10·84
Propane	39	16.0	76·4	16·67
Butane	38	16·0	69·5	17·10
Carbon monoxide	43	18·0	71·2	0·495
Ethylene	70	29·6	192·5	7·58

[a] further discussed on page 47.
[b] further discussed on page 46.
[c] further discussed on page 50.

3.4   COMPLETENESS OF COMBUSTION

Reduced to its simplest form the combustion of hydrocarbons in a flame amounts to the following. In the course of the combustion process fuel molecules are preheated and subsequently react with oxidant molecules to produce combustion products, mainly water and CO_2. Simultaneously, certain thermal cracking reactions take place which result in the formation of carbon, unsaturated compounds and polymers.

In the interest of clean combustion, high efficiency and minimum maintenance it is essential that such side reactions leading to carbon and unsaturates should be largely eliminated and that combustion to carbon dioxide and water should predominate. The design of gas burners, appliances and combustion systems is generally such that gases of a given tendency towards cracking and decomposition can be effectively burnt without undue carbon formation. It is, however, essential that the carbon forming tendency of any substitutes should not exceed that of the standard supply.

The tendency of hydrocarbon gases to produce unsaturates or soot before or in preference to being oxidised to clean combustion product is expressed in terms of such parameters as 'yellow-tipping index', which is a measure of yellow-burning aromatics and olefins present or formed in the gas, and 'sooting index', a parameter indicating soot-forming tendency.

3.5   ESTABLISHED INTERCHANGEABILITY SYSTEMS

The purpose of gaseous interchangeability standards is to ensure that a basic gas supply is matched by any replacement in regard to all its relevant combustion properties. The fact that these properties do not stand on their own but can influence each other has led to two results: firstly, the choice of possible combustion parameters is rather wider and a number of different interchangeability systems have been developed. Secondly, limiting values of any parameter may not be absolute but often depend on the value of other parameters. The interchangeability criterion thus becomes an area in a graph—if two interdependent characteristics are limiting—or even a volume in space if there are three or more restrictive parameters that are mutually interdependent.

It is proposed to discuss briefly in subsequent sections some of the more widely accepted interchangeability systems that are being used in Europe and in the United States. In doing so it should be borne in mind that the limiting values for each parameter are empirical, they amount to the tolerances of the most sensitive burners and appliances in a particular population. They can, in consequence, change as certain equipment becomes obsolete. Less frequently additional restrictions may result from the introduction of new equipment into a particular distribution area.

Because of the empirical character of these rules and their strict applicability in certain markets only, it is not altogether surprising that different conclusions may be arrived at, depending on which set of rules is applied. In other words two gases which, according to one system should not be interchangeable, may occasionally, when applying a different set of rules, be found to be fully exchangeable after all. It is, therefore, sometimes desirable to compare substitutes with their basic standards using more than one of the following groups of parameters.

3.5.1   The Delbourg System[2,3]

This is one of the most carefully analysed sets of rules for gases of both the first (coal gas type) and the second family (natural gases) and applies specifically to burners and appliances on the French market. Gases are said to be interchangeable if they fall within a triangle on a 'combustion potential' versus modified Wobbe Index plot.

The triangle is formed by three lines representing:
the transition from safe to non-hygienic combustion, characterised by the presence of more than 0·5 % vol of carbon monoxide
the change from stable flames to lifting and blow-off conditions, and
the occurrence of flash-back inside the burner.

This triangle is determined empirically, on the one hand, by means of a test burner (the 'brûleur contrôleur' of Gaz de France); on the other hand, by means of a large number of burners and appliances, all adjusted to burn the reference gas and tested by means of mixtures of test gases for flame lifting, carbon monoxide formation and flash-back. A typical Delbourg diagram is shown in Fig. 3.2.

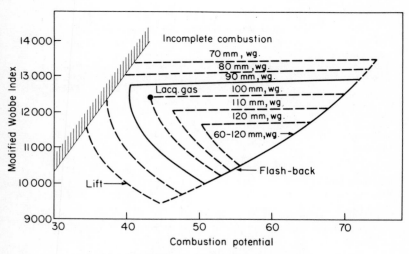

FIG. 3.2   Delbourg interchangeability diagram.

The co-ordinates of the diagram, the modified Wobbe Index and combustion potential, are unique characteristics of the fuel gas under test. The modified WI is obtained by multiplying the gross calorific value based WI by the two modifying factors mentioned on page 38, one to allow for the presence of higher molecular weight hydrocarbons, the other to take care of oxygen present in the gas either as such or in the form of carbon dioxide. (Both these corrections are insignificant for gases consisting mainly of methane.)

The other co-ordinate, the combustion potential, is an indirect measure of flame speed. It is based on a series of factors that are typical of each gas component *viz.* hydrogen, carbon monoxide, methane, hydrocarbons and inert gases. The Delbourg combustion potential (c.p.) is calculated with the aid of the formula

$$c.p. = \frac{k_1 k_2 \sum ia^*}{\sqrt{d}}$$

where i is the volume percent of each component, a the corresponding factor and d the specific gravity. The numerical values of a for different gases are listed in Table 3.3.

TABLE 3.3
Calculation of Burning Velocity Parameters

Interchangeability system: Parameter:	Delbourg Combustion potential a	Weaver Gilbert-Prigg Flame speed F	Holmqvist Characterisation factor X_K
Hydrogen	1·0	339	0·291
Methane	0·3	148	7·24
Ethane	0·75	301	10·59
Propane	0·95	398	13·34
Butanes	1·0	513	14·05
Carbon monoxide	0·7	61	0·487
Ethylene	1·9	454	7·66
Calculation of parameters for a gas mixture:	$\dfrac{k_1 k_2 \sum ia}{\sqrt{d}}$	$\dfrac{\sum iF}{\sum iA + 5I - 18\cdot8\, O_2 + 1}$	$\dfrac{\sum i X_K}{\sqrt{d}}$

The Delbourg system is used widely in French speaking countries, Italy and elsewhere. It is probably more effective in interpreting the interchangeability of natural gases than almost any other systems, but it has been claimed that it is too restrictive and designed around

---

* The constants k_1 and k_2 are empirical and are functions of the oxygen content and the modified Wobbe Index. For most natural gases their value = 1·0.

French appliances of very limited flexibility, and is furthermore, as shown in Fig. 3.2, very much a function of gas pressure.

In addition to modified WI and combustion potential the Delbourg system also assesses yellow-tipping and sooting tendency using appropriate factors for each gas component. The total yellow-tipping or sooting index equals the sum of the mole percentages of the component multiplied by the corresponding factor. For genuine interchangeability neither yellow-tipping nor sooting tendency must exceed that of the reference gas.

3.5.2 The Gilbert-Prigg System[5]

This is another system using graphical interpretation. The variables plotted in a Gilbert-Prigg graph are Wobbe Index and Weaver flame speed factor. The resultant interchangeability diagram is triangular with boundaries formed by lines of incomplete combustion, flash-back and flame failure by blow-off.

The diagram is widely used in the United Kingdom and other Commonwealth countries; the limits themselves have been tested by means of the so-called 'aeration test burner', and UK burners and appliances have for many years been designed for a series of reference gases of different Wobbe Indices. Flexibility of burners and appliances had to be such that they would be satisfied by any gas within the range established by the ATB.

While some problems have arisen when trying to apply Gilbert-Prigg criteria to gases of the second family—the diagram was very effective in conjunction with coal gas and naphtha-derived replacement gases—these appear to have been overcome and the system is now used for both first and second family gases.[6]

Figure 3.3 shows a typical Gilbert-Prigg interchangeability area for a UK town gas (G4) and also indicates the location of the various types of gas components in the diagram. The Wobbe Index (unmodified) is plotted as ordinate and the Weaver flame speed factor of the gas as the abscissa.

The latter is not a simple burning velocity but a relative rate, with hydrogen assigned 100—actual velocities can be measured in a number of different ways and depend on such extraneous factors as pressure, temperature and humidity. Flame speed factor

$$S = \frac{\sum F_1}{\sum A_1 + 5I + 18 \cdot 8\, O_2 + 1}$$

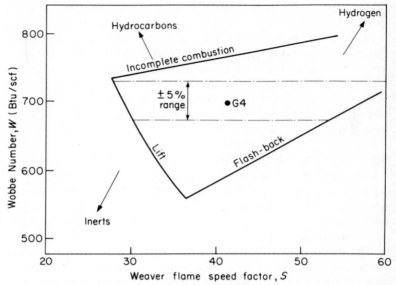

FIG. 3.3   Gilbert-Prigg interchangeability diagram (G4 gases).

where A_1 is the stoicheiometric air requirement of each component,
multiplied by its mole fraction,

F_1 is the Weaver flame speed factor of the component
multiplied by its mole fraction,

I is the mole fraction of inerts,

O_2 is the fuel oxygen content

Values of F have been published and are listed in Table 3.3.

3.5.3 The Weaver System[12]

As distinct from the Delbourg and Gilbert-Prigg systems, the Weaver
criteria for interchangeability are totally independent of each other
and according to Weaver there is, therefore, no need for graphical
representation or an interchangeability diagram.

Weaver defines five indices:

The lifting index $\qquad\qquad J_L = J_r \times \dfrac{S_s}{S_r} + \dfrac{(100 - O_2)_s}{(100 - O_2)_r}$

The flash-back index $\qquad\quad J_F = \dfrac{S_s}{S_r} - 1{\cdot}4\,J_r + 0{\cdot}4$

The yellow-tipping index $\qquad J_y = J_r + \dfrac{N_s - N_r}{110} - 1$

The CO-forming index $\qquad J_1 = J_r - 0.366\dfrac{R_s}{R_r} - 0.634$

The heat-output index $\qquad J_H = \dfrac{(WI)_s}{(WI)_r}$

and explains that subscripts s and r refer to substitute and reference gases; S is flame speed, J_r is the change in primary aeration due to substitution and N and R are factors characteristic of soot forming tendency and molecular unsaturation.

For complete interchangeability it would be desirable that heat output and lifting indices should be 1 and flash-back, yellow-tipping and CO-forming indices zero. Divergences of up to 5% in either direction are generally considered permissible.

3.5.4 The Use of Knoy's Formulas[8,9]

Since non-hydrocarbon components are rare in US natural gas supplies, many US utilities are only concerned with heat output from a burner, and other gas characteristics that do not affect thermal throughput can in their view be neglected.

The use of the two formulas:

$$C = (CV - 175)/\sqrt{d} \text{ with permissible variations of } \pm 5\%$$

and

$$P_s = CV_s \sqrt{d_r} \bigg/ \left(\frac{\sqrt{d_r} + CV - P_r}{P_r} \sqrt{d_s} \right) \text{ which can vary over} \pm 15\%$$

is based on the concept that in a well designed burner P, the heating value of the gas mixture at the burner head, equals 175 Btu/ft^3. If CV is the calorific value of the gas then C becomes the difference in Wobbe Index between the undiluted and the aerated gas, and Knoy asserts that this should normally not vary by more than $\pm$ 5%.

The second formula, in which subscripts s and r indicate substitute and reference gases, permits one to work to values other than $P = 175$ and to adjust both heating value CV and specific gravity d of the substitute gas in such a way as to keep P_s within the specified bounds of $\pm$ 15%.

Flame speed, and therefore flame shape, sooting tendency and yellow tipping are not considered important variables.

3.5.5 Other Interchangeability Systems

Apart from the Weaver and Knoy assessments of gaseous interchangeability, the American Gas Association has also published its method.[1] In Europe, an alternative to the UK and French systems has been claimed by Holmqvist in Sweden who uses C_K, the so-called characterisation factor as a flame speed parameter.[7]

$$C_K = \frac{k_1 \sum i X_K}{\sqrt{d}}$$

where k is a function of the inert content of the gas mixture, i is the mole fraction of the component, X_K the appropriate factor listed in the last column of Table 3.3, and d the specific gravity of the gas.

The other parameter used by Holmqvist is the Wobbe Index of the gas based on its net calorific value. For complete interchangeability $WI_{net.}$ should not vary by more than $\pm 5\%$. Variations in C_K are less carefully defined but should probably not exceed $\pm 3\%$.

3.6 THE RELATIVE MERITS OF DIFFERENT INTERCHANGEABILITY SYSTEMS[11]

The purpose of setting up interchangeability standards and appropriate tolerances for gas quality parameters is to ensure that gas burners, domestic appliances and industrial equipment operate satisfactorily on all types of gases that are likely to be distributed in a given district. Permissible variations in gas quality are thus closely linked to the flexibility of gas burners in use, particularly those of domestic appliances. The latter are considered most critical, not only because they have to meet very accurate performance standards such as oven temperature, simmering setting of cooker plates, drying of laundry to precise humidity levels, etc., but also because, unlike industrial burners, they are rarely checked and reset by specialist personnel.

Permissible tolerances or interchangeability ranges thus become a function of the type of domestic appliances in a particular area; in certain instances it can actually be one or two critical appliances in use in a large number of homes that determine certain quality aspects. A point to note, incidentally, is that manufacturers of appliances work to established gas quality standards; once these are set up, on the strength of the existing appliance population, new appliances will be designed to match existing gas quality and its variations. In other words, interchangeability standards become not only firmly established but also self-propagating.

It is, therefore, not at all surprising to find that different interchangeability systems have developed in different countries, particularly since the only opportunity to arrive at uniform standards, the transition from coal gas (1st family) to natural gas (2nd family), was not used to establish world-wide quality standards for both gases and appliances. This was mainly because the transition occurred at different times in different countries, and also because there remained quality differences even within each family of gases.

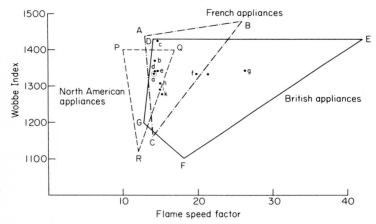

FIG. 3.4 Comparison of interchangeability prediction methods. ⊙ a, reference gas—North Sea A; • b, US natural gas; c, Algerian LNG; d, CRG gas—double methanation; e, CRG gas—hydrogasification; f, FBH of crude oil-methanation; g, GRH—CO_2 removed, C_3H_8 enriched; h, HYGAS; i, BI-GAS; k, CO_2 acceptor.

Figure 3.4 is indicative of these differences: it shows, with the aid of a Wobbe Index v. flame speed factor diagram, the area of interchangeability of the same reference gas using French, North American and British tolerances. The triangle ABC thus shows what would be acceptable variations in quality for appliances in France and the area DEFG indicates the corresponding flexibility of UK equipment. US gas quality tolerances, expressed by the triangle PQR, are narrower and indicative of less flexible appliances than those marketed in Europe, especially in regard to flame speed.

It would seem, therefore, that the relative merits of the different interchangeability systems are hard to assess, as they are so closely tied to particular appliance populations. One can, however, generalise in at least one respect as far as Holmqvist, Gilbert-Prigg and Delbourg are concerned: the two triangles are stretched horizontally, *i.e.* cover a wide range of burning velocities, wider in fact than one usually encounters in those pure hydrocarbon gases that contain neither hydrogen nor olefins. The flame speed, combustion potential and Holmqvist characterisation number thus become no longer critical and can, in our opinion, normally be disregarded. This means that interchangeability, at least in Europe, apart from soot forming or yellow-tipping tendency, can be interpreted as being the permissible variation in Wobbe Index (standard WI for Weaver and Gilbert-Prigg; modified WI for the Delbourg, and WI based on net calorific value for the Holmqvist system).

Variations in heat load are generally considered acceptable if they do not exceed $\pm 5\%$, and there is reasonable agreement in this respect between the four authorities mentioned above. A simplified but nevertheless generally useful standard requirement for all forms of SNG in Europe would therefore be that their Wobbe Indices (standard and derived functions) should not differ from those of the original gas supply by more than 5%.

As for US appliances, it should be noted that, at least in theory, these do not permit corresponding variations in flame speed and that they are therefore less flexible in regard to gas composition. It is believed, however, that the narrow range specified by Weaver is due more to the general absence of natural gases of different types than to an inherent lack of flexibility of US made appliances. It follows that the combustion properties of a substitute gas will have to be more closely defined and are thus more restrictive in some gas markets than in others.

3.7   APPLICATION OF INTERCHANGEABILITY DIAGRAMS TO SNG

Typical characteristics for gases made by the different routes to be described in later chapters are listed in Table 3.4 and Appendix A, and their location on the diagram in Fig. 3.4 (c to k) is an indication of whether or not they would be considered *a priori* acceptable for either

TABLE 3.4
Typical Combustion Properties of Natural and Substitute Gases

Gas	Described in Chapter	Point on diagram (Fig. 3.4)	Wobbe Index Btu/ft³	Weaver flame speed factor
North Sea Gas	2	a	1335	14·2
US Natural Gas	2	b	1370	14·3
Algerian LNG	2	a	1425	14·8
LT Reformed				
CRG-Double methanation	6	d	1342	14·2
CRG-Hydrogasification	6 + 7	e	1342	14·4
Fluid bed hydrogenated crude				
(after methanation)	7	f	1335	19·8
Gas recycle hydrogenation				
(CO₂ removed, C₃H₈ added)	7	g	1342	26·2
HYGAS	9	h	1310	15·0
BI-GAS	9	i	1290	14·9
CO₂ Acceptor	9	k	1275	15·3

complete substitution or only partial replacement of the base gas (a or b). It is worth noting that low temperature steam reforming products (d, e) and coal derived substitute gases (h, i, k) come closest to full interchangeability and that hydrogenation produces gases (f, g) of acceptable Wobbe Index but too high a flame speed factor, especially as far as French and US appliances are concerned. We would like, however, to repeat our previously stated reservations as to the real significance of these differences, particularly since there appear to be no variations in Wobbe Index flexibility between the three appliance populations.

In conclusion, it is felt that the standard Wobbe Index or one of the derived functions capable of predicting thermal output remains the simplest and probably the most reliable practical method of assessing interchangeability.

While graphical approaches and analytical formulas that would permit one to allow for differences in burning velocity and for the interaction between Wobbe Index and flame speed factor undoubtedly have their uses, it is nevertheless true that they can lead to contradictory results, which then have to be explained away by alleged differences between appliance populations.[4] It is, therefore, both intellectually and in practice, easier to disregard these differences and to confine our attention, at least as far as hydrocarbon gases are concerned, mainly to the Wobbe Index criterion.

REFERENCES

1. American Gas Association (1950). *Interchangeability of Various Fuel Gases With Manufactured Gases*, Res. Bull. 60, Cleveland, Ohio.
2. Delbourg, P. (1951). *Le contrôle de la qualité et l'interchangeabilité des gaz*, Congrès de l'Ass. Tech. de l'Indus. du Gaz.
3. Delbourg, P. and Schneck, H. (1957). *Interchangeabilité des gaz de la deuxième famille*, Congrès de l'Ass. Tech. de l'Indus. du Gaz.
4. France, D. H. (1974). U.S. and U.K. substitute gas supplies, *Energy World*, No. 11, 10–13.
5. Gilbert, M. G. and Prigg, J. A. (1956). *The Prediction of the Combustion Characteristics of Town Gas*, Gas Council Res. Comm. C 35.
6. Harris, J. A. and Lovelace, D. E. (1968). *J. Inst. Gas Engrs*, **8**, 169.
7. Holmqvist, R. (1957). *Svenska Gäsforeningens Arsbok*, 35–80.
8. Knoy, M. F. (1941). Combustion experiments with liquid petroleum gases, *Gas*, **17**, 14–19.
9. Knoy, M. F. (1947). Master interchangeability chart, *Gas*, **23**, 46–52.
10. Rosenberg, R. B. and Weil, S. A. (1973). *Interchangeability of Imported Natural Gas and SNG*, Paper 24, Inst. Gas Tech., SNG Symposium I.
11. Theron Mulder, J. C. (1961). *A Comparative Study of Gaseous Interchangeability Systems*, Paper 4, Int. Gas Congress, Stockholm.
12. Weaver, E. R. (1951). Formulas and graphs for representing the interchangeability of fuel gases, *J. Res. Nat. Bur. Standards*, **46**, 213–45.

Chapter 4

Feedstocks for SNG Manufacture

4.1 INTRODUCTION

The basic reactions underlying the conversion of fossil fuels into SNG, as discussed in Chapter 5, are the addition of hydrogen to, or the removal of some of the carbon from a more complex hydrocarbon molecule. The molecular structures that can be hydrogenated or decarbonised cover a wide range of molecular weights and structural complexity, and the feedstock may be contaminated to a varying extent by non-hydrocarbons. Different feed preparation, feed purification and processing routes must therefore be adopted if these widely differing materials are to be converted into one and the same type of end product.

The present chapter is devoted to a discussion of the properties of SNG feedstocks, including solid fuels such as the various grades of coal and lignite, coke and anthracite; liquid petroleum fuels, *i.e.* crude oil and the fractions obtained from crude by conventional petroleum refining processes; a range of liquid products sometimes produced during the purification of natural gas, namely condensate comprising propane, butanes and so-called natural (or casing head) gasoline (*see* also Chapter 2).

It would clearly be impracticable in the present context to review all the properties of the above mentioned range of materials. It is therefore proposed to deal mainly with those aspects that have a bearing on the further processing of the feedstocks. Brief mention will be made in connection with each feedstock of conversion routes that would be used to produce SNG from such a raw material. However the SNG processes themselves will be described in later chapters.

4.2   COAL AND OTHER SOLID FUEL FEEDSTOCKS

4.2.1 Gasification Characteristics

The conversion of a solid fuel into a gas involves a number of steps of which the most important are comminution, feed preparation, preheating, interaction with a gaseous reactant, chemical reaction of hydrocarbons with steam, hydrogen and oxygen, formation of gaseous products, gas purification and send-out. Certain properties of coal feedstocks have a more immediate effect on their performance in gasification processes than others, among them the following:

Calorific Value: the thermal energy contained in the feedstock will emerge, after certain process losses, in the principal gaseous product. Feedstocks of a high calorific value therefore normally have a premium value.

Volatiles Content: the primary reaction is generally a simple heat treatment that separates volatiles and char. The former are already higher in hydrogen content and therefore more easily converted into SNG. A high volatiles content is thus basically desirable.

Coking Tendency: coal particles when heated swell, exude tar and tend to adhere to each other. This will interfere with the smooth operation of fluidised beds, the transport of heated coal in powder form and its storage. Coking coals are less suitable for gasification unless they can be pretreated, usually by some form of pre-oxidation step, to reduce their caking and swelling tendency.

Ash Content: coal feedstocks should have a low ash content in order to minimise unnecessary processing and energy losses. Coal feedstocks are, as a rule, prepared by cleaning and washing. Provided ash is inherently low or the impurities can be removed by a preliminary treatment, the coal in question becomes acceptable for gasification.

Moisture Content: all coals contain moisture and may have to be dried before processing. Some are washed in order to reduce ash and impurities, and end up with a high moisture content. The latter is, therefore, not inherently undesirable, but in order to ensure easy processing water and moisture should be distributed evenly.

Sulphur and other Contaminants: most coals contain some sulphur and concentrations of up to 4 and 5%wt have been found occasionally. Some of the sulphur, *e.g.* that due to the presence of pyrites, can be removed by washing and other methods of feedstock preparation; the remainder is chemically linked and is therefore only released, as a rule in the form of H_2S, when the bulk of the coal is gasified. Since H_2S must ultimately be removed from the gas and may also interfere with certain processing aspects, particularly catalytic reactions, a high sulphur content is inherently undesirable.

Coals also contain some nitrogen and oxygen. While the presence of the latter does not interfere with SNG routes where part of the coal is oxidised, it is undesirable in hydrogenation processes since it results

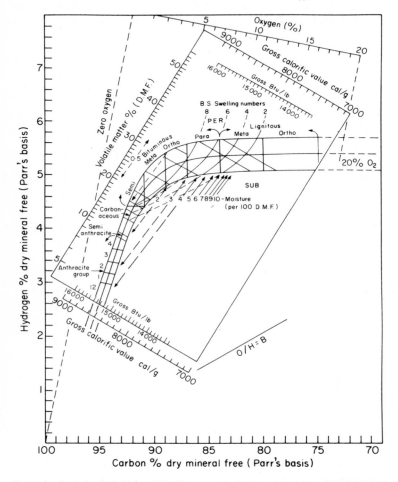

FIG. 4.1   Seyler's Coal Chart 47B. (Courtesy: Brit. Nat. Committee, World Energy Conf. London.)

in additional hydrogen consumption. A high nitrogen content in the fuel results in a finished gas that will be diluted with inerts or ammonia and will therefore tend to be of a lower calorific value. This will be undesirable unless other high calorific value components are present to balance the deficiency.

Typical values for the coal characteristics mentioned above are listed in Table 4.1, which compares bituminous coals, anthracites and cokes and lists their range of compositions, volatiles contents and calorific values, first on a dry and ash-free, then on an 'as found' basis.

TABLE 4.1
Typical Properties of Solid Fuels

	Bituminous coals[a]	Anthracites[b]	Coke
Dry and ash-free basis			
Analysis, %wt			
Carbon	81·0–89·5	92·5–94·5	96·5
Hydrogen	5·3–4·9	4·0–3·0	0·5
Nitrogen	1·9–1·5	1·4–1·1	} 2·0
Sulphur	1·0–0·6	0·9–0·7	
Oxygen	11·5–2·1	1·3–0·9	1·0
Calorific value (gross)			
Btu/lb	14 000–15 500	15 000–15 800	15 000
kcal/kg	7 778–8 612	8 334–8 778	8 334
Volatile matter, %wt	40–25	11·5–5·0	1·0
As found basis (moisture about 7%, ash about 8%)			
Analysis, %wt			
Carbon	59·0–75·5	76·8–78·2	82·0
Hydrogen	3·6–4·3	2·4–3·8	0·4
Nitrogen	1·2–1·7	0·9–1·2	} 1·7
Sulphur	1·2–1·7	1·0	
Oxygen	7·6–7·8	1·5–2·2	0·9
Calorific value (gross)			
Btu/lb	10 300–12 700	12 800–13 900	12 300
kcal/kg	5 723–7 056	7 112–7 723	6 834
(net) Btu/lb	9 710–11 790	12 440–12 820	12 190
kcal/kg	5 394–6 551	6 912–7 123	6 773

[a] The bituminous range includes coking, weakly coking and strongly coking types.
[b] Including semi-anthracites.
Source: Spiers, H. M., *Technical Data on Fuel*, World Power Conference, London, 1962.

The table shows that the properties of coal are closely linked with 'coal type' and a discussion of this concept would therefore seem opportune. Several attempts to classify coals according to rank (*i.e.*

TABLE 4.2
The Coal Classification System Used by the British National Coal Board

Group (S)	Class	Volatile matter dry, mineral matter-free %	Gray-King coke type[a]	General description
100		Under 9·1	A	
	101[b]	Under 6·1	A	Anthracites
	102[b]	6·1–9·0		
200		9·1–19·5	A–G8	Low-volatile steam coals
	201	9·1–13·5	A–G	
	201[a]	9·1–11·5	A–B	Dry steam coals
	201[b]	11·6–13·5	B–C	
	202	13·6–15·0	B–G	
	203	15·1–17·0	B–G4	Coking steam coals
	204	17·1–19·5	G1–G8	
	206	9·1–19·5	A–B for V.M. 9·1–15·0 A–D for V.M. 15·1–19·5	Heat-altered low-volatile steam coals
300		19·6–32·0	A–G9 and over	Medium-volatile coals
	301	19·6–32·0	G4 and over	
	301[a]	19·6–27·5	G4 and over	Prime coking coals
	301[b]	27·6–32·0		
	305	19·6–32·0	G–G3	(Mainly) heat-altered
	306	19·6–32·0	A–B	medium-volatile coals
400–900		Over 32·0	A–G9 and over	High-volatile coals
400		Over 32·0	G9 and over	
	401	32·1–36·0	G9 and over	Very strong caking coals
	402	Over 36·0		
500		Over 32·0	G5–G8	
	501	32.1.3		Strongly caking coals
	502	Over 32.1.0	G5–G8	
600		Over 32·0	G5–G4	
	601	32·1–36·0		Medium caking coals
	602	Over 36·0	G1–G4	
700		Over 32·0	B–G	
	701	32·1–36·0		Weakly caking coals
	702	Over 36·0	B–G	
800		Over 32·0	C–D	
	801	32·1–36·0	C–D	Very weak caking coals
	802	Over 36.0		
900		Over 32·0	A–B	
	901	32·1–36·0	A–B	Noncaking coals
	902	Over 36·0		

[a] Coals of groups 100 and 200 are classified by using the parameter of volatile matter alone. The Gray-King coke types quoted for these coals indicate the ranges found in practice and are not criteria for classification.

[b] To divide anthracites into two classes, it is sometimes convenient to use a hydrogen content of 3·35% (d.m.m.f.) instead of a volatile matter of 6·0% as the limiting criterion. In the original Coal Survey rank coding system the anthracites were divided into four classes then designated 101, 102, 103, and 104. Although the present division into two classes satisfies most requirements, it may sometimes be necessary to recognise four or five classes.

Courtesy: National Coal Board.

TABLE 4.3
ASTM Classification of Coal by Rank[a]

Class	Group	Limits of fixed carbon Btu, mineral-matter-free basis	Requisite physical properties
Anthracitic	1. Meta-anthracite	Dry FC, 98% or more (dry VM, 2% or less)	
	2. Anthracite	Dry FC, 92% or more and less than 98% (dry VM, 8% or less and more than 2%)	
	3. Semi-anthracite	Dry FC, 86% or more and less than 92% (dry VM, 14% or less and more than 8%)	Nonagglomerating[b]
Bituminous	1. Low-volatile bituminous coal	Dry FC, 78% or more and less than 86% (dry VM, 22% or less and more than 14%)	
	2. Medium-volatile bituminous coal	Dry FC, 69% or more and less than 78% (dry VM, 31% or less and more than 22%)	
	3. High-volatile A bituminous coal	Dry FC, less than 69% (dry VM, more than 31%); and moist[c] Btu, 14 000[e] or more	
	4. High volatile B bituminous coal	Moist[c] Btu, 13 000 or more and less than 14 000[e]	
	5. High-volatile C bituminous coal	Moist Btu, 11 000 or more and less than 13 000[e]	Either agglomerating or non-weathering[f]
Sub-bituminous	1. Sub-bituminous A coal	Moist Btu, 11 000 or more and less than 13 000[c]	Both weathering and non-agglomerating
	2. Sub-bituminous B coal	Moist Btu, 9 500 or more and less than 11 000[e]	
	3. Sub-bituminous C coal	Moist Btu, 8 300 or more and less than 9 500[e]	
Lignite	1. Lignite	Moist Btu, less than 8 300	Consolidated
	2. Brown coal	Moist Btu, less than 8 300	Unconsolidated

(FC = fixed carbon; VM = volatile matter; Btu = British thermal units)
[a] This classification does not include a few coals that have unusual physical and chemical properties and that come within the limits of fixed carbon or Btu of the high-volatile bituminous and sub-bituminous ranks. All these coals either contain

age), volatiles content, coking tendency or chemical composition have been made over the years.

In one of the earliest attempts, the well known Seyler Chart reproduced in Fig. 4.1, a plot of hydrogen *versus* carbon content, both on a dry-and-ash-free basis, shows that practically all coals from anthracites to lignites fall into a narrow band. Superimposed grids allow one to read off and predict gross calorific value or volatiles content, oxygen content and coking tendency in terms of British Standard Swelling Numbers.

Other coal classification systems are in use in the UK (the National Coal Board system) and in the US (the ASTM Classification by Rank), the latter being also widely accepted in other countries. Both classify coals mainly according to their volatile contents. In the US a subsidiary criterion is calorific value of the coal as found. In the UK caking and coking properties of the coal are also defined. The latter is therefore of direct use in assessing gasification performance in different SNG processes. The two systems are summarised in Tables 4.2 and 4.3.

Other ways of characterising coal are based on its mineralogical properties. While seemingly uniform, coal is in fact made up of several mineral components which differ in physical and chemical properties. The principal components have been termed vitrain, fusain and clarain and their relative proportions differ in coals of different rank, coking tendency and volatiles content, although no simple and readily understandable correlation so far appears to have been established.

4.2.2 Occurrence of Coals for Gasification

Anthracites, semi-anthracites, bituminous coals, semi-bituminous coals, lignites and various other solid hydrocarbon fuels have been

less than 48 % dry, mineral-matter-free fixed carbon or have more than 15 500 moist, mineral-matter-free Btu.

[b] If agglomerating, classify in low-volatile group of the bituminous class.

[c] Moist Btu refers to coal containing its natural bed moisture but not including visible water on the surface of the coal.

[d] It is recognised that there may be noncaking varieties in each group of the bituminous class.

[e] Coals having 69 % or more fixed carbon on the dry, mineral-matter-free basis shall be classified according to fixed carbon, regardless of Btu.

[f] There are three varieties of coal in the high-volatile C bituminous coal group: variety 1, agglomerating and nonweathering; variety 2, agglomerating and weathering; variety 3, nonagglomerating and nonweathering.

Reprinted from copyright protected material published by ASTM.

found in many parts of the world. Interest in coal arose at the beginning of the industrial revolution when coke replaced charcoal as a reducing agent for iron ores. Coking coals of low ash content were thus primarily in demand and weakly coking or non-coking coals, while widely required for steam raising and domestic applications, were of somewhat lower interest. In today's markets coking coals are still at a slight premium, although the price differential has shrunk in percentage terms, and there is little or no incentive to use these grades for SNG manufacture.

We can also eliminate anthracites and semi-anthracites from consideration as SNG feedstocks since they are low in volatiles, tend to be unreactive, are in relatively short supply and generally have a ready market in domestic and commercial central heating systems. Finally, one can probably discount very low grade lignites because of their high water and ash contents and relatively low calorific value.

This means that most coal-based gas and SNG plants will tend to use bituminous, semi-bituminous and lignitic non-coking coals, *i.e.* basically the same grades that today find their main market in steam raising and electricity generation. In the light of this competition it seems clear that coal gasification will be of particular interest only where there are large reserves of steam coal, far from electricity and industrial sales outlets. It seems to follow that, in spite of the

TABLE 4.4
Coal Reserves and Production, World-wide

Country	Reserves metric tons × 10^6	Production, 1970 metric tons × 10^3
USA	1 100 000	550 388
Canada	61 000	11 598
USSR	4 121 603	432 715
Europe		
United Kingdom	15 500	144 564
Belgium	1 796	11 362
France	2 800	37 364
Italy		295
West Germany	70 000	111 443
Netherlands	2 394	4 334
China	1 011 000	360 000
Japan	19 248	39 694
World	6 641 200	2 124 000

Source: *Statistical Year Book*, United Nations, 1972.

widespread occurrence of coal, not many deposits will be used for gasification in the near future.

Latest coal reserves and coal production figures for those countries of the world where appropriate statistics are available are summarised in Table 4.4 which shows that both production and reserves are highest in the US, Russia, and China, where it will take many years to exhaust available reserves. Western European demand, on the other hand, outstrips available supplies and substantial quantities will have to be imported from the surplus areas listed and also from South Africa and Australia.

It does not appear likely, in consequence, that coal gasification will be introduced into Europe in the near future. Similarly Russia, because of its large reserves of natural gas, or China, in the light of its relatively low level of industrialisation, will not gasify their coal. The United States are, in fact, the only gas market where coal gasification plants are under construction and new coal processes are being developed.

4.3 PETROLEUM FEEDSTOCKS

The manufacture of SNG from petroleum products can be based on almost any petroleum fraction or on unfractionated crude oils of a wide range of origin and quality. While catalytic steam reforming reactions usually require sulphur-free feedstocks and are, therefore, restricted to light fractions that can be easily purified, other routes such as partial oxidation operate without a catalyst and will accept even the heaviest petroleum feedstocks. Hydrogenation processes, depending on whether catalysts are present and on their sensitivity to contamination, may be restricted to light products or may tolerate distillates, crude oils and in some instances heavy fuel oil feedstocks. This implies that the entire range of petroleum fuels must be considered potential SNG feedstocks.[2,3]

4.3.1 Crude Oils
The bulk of petroleum products marketed around the world are derived from crude petroleum, a mixture of liquid and sometimes gaseous hydrocarbons produced from oil-bearing rocks in many parts of the world. The main oil-producing areas with their 1974 rate of production and their remaining oil reserves are listed in Table 4.5.

Again these figures must be viewed critically; production rates can change from year to year, and the definition of reserves, whether proven, probable or potential, is not always clear; new discoveries are made and older figures may have to be revised in the light of recent information.

However, there is no doubt that certain regions of the world will have surpluses and will continue to export (Middle East, North

TABLE 4.5
Oil Production and Reserves of Some Oil-producing Areas (million tons)

	Proven and forecasted reserves	Annual production	
		1970	1974
North America	6 660	492	592
Middle East	56 670	684	1 094
Iran	—	196	301
Saudi Arabia	—	181	412
Others	—	307	381
Venezuela, Trinidad	3 630	190	165
Algeria, Libya	5 000	209	130
Nigeria	900	54	112
Indonesia	1 900	51	72
Australia	240	4	18
USSR	10 000	333	457
China	?	3	65
World	85 000	2 272	2 870

Sources: *Statistical Yearbook*, United Nations, 1973; *Petroleum Economist*, January 1975.

Africa, Nigeria, Indonesia and, for some time, Venezuela); others will be roughly in balance (Russia, Australia and hopefully the US). Finally, some regions such as Western Europe and Japan will remain importers.

The quality of crude oils varies between wide limits. As shown in Table 4.6, the lightest, usually paraffinic crudes are made up of saturated hydrocarbons and contain up to 85% vol distillates. Very heavy, generally asphaltic crudes, on the other hand, may produce as little as 20% vol distillates.

While gasification efficiency and gasification economics are functions not only of feedstock quality but also of conversion route and detailed process design, there is little doubt that light, low sulphur, paraffinic crudes perform better in most SNG processes and are cheaper to convert than are heavy sour asphaltic crudes. In particular, the lighter materials require less hydrogen transfer and less carbon removal; they form fewer refractory aromatic by-products; they contain less sulphur and other catalyst poisons and, in the course of gasification, they form less H_2S and CO_2, which subsequently have to be removed from the gas at extra cost.

TABLE 4.6
Range of Properties of Crude Oils

	Light paraffinic crude	Heavy asphaltic crude
Composition % wt		
Carbon	83·9	86·8
Hydrogen	14·0	11·4
Sulphur[a]	0·06	2·0
Nitrogen	0·11	1·7
Oxygen	up to 0·5	
Metals (Fe, V, Ni, etc.)	up to 0·03	
Specific gravity, 15/15 °C	0·77	1·0
Distillate yield, % vol	65–85	20–30

[a] Widely variable; 0·01–3 approx.

Typical crude oils that lend themselves to relatively efficient conversion are light North Sea oil, light Algerian, light Arabian and light Iranian (e.g. Aga Jari) in that order of preference. Less suitable for gasification are heavy oils such as Kuwait, heavy Venezuelan (e.g. Tia Juana), heavy Arabian and Sicilian (Gela) crudes.

4.3.2 Refining of Crude Oil[4]
The complex make-up of crude oil and the fact that it consists of literally hundreds of chemical compounds whose physical properties and chemical reactivities vary between wide limits, puts a considerable premium on processes that will subdivide this multitude of products into groups of more or less uniform properties, which can then be processed much more effectively under optimal conditions. A series of chemical engineering processes, collectively known as petroleum refining, are in fact used to convert crude oil into products of uniform and both nationally and internationally standardised properties.

While the main purpose of oil refining in the past has been the manufacture of liquid fuels carefully tailored for subsequent use in industrial furnaces, domestic heating systems, diesel engines, turbojet engines, and especially spark-ignition engines, another aim, the manufacture of chemical feedstocks, which has much more in common with gasification feedstock production, has recently become of considerable importance. Thus, in addition to producing motor spirit, the special properties and high price of which justified complex processes such as catalytic cracking and catalytic reforming, modern oil refineries also produce primary (straight-run) distillates of various boiling ranges that can be used either for subsequent conversion into petroleum chemicals or as feedstocks for the manufacture of gases such as SNG.

The straight fractionation of crude petroleum results in a series of distillates of standard boiling range, irrespective of crude origin although relative yields of different products vary from one crude oil to another. These products can be used as such for certain applications, including chemical/gasification conversion, or may be further processed. Thus removal of the most volatile products which boil below 35 °C under atmospheric pressure yields LPG; the next heavier fraction, 35–200 °C is the basis of gasoline production but can also be subdivided into one or two 'naphthas' for chemical and gasification feedstocks. Kerosines for aviation turbines and domestic burners equipped with wicks boil between 150 and 300 °C, middle distillates for use in high speed diesel engines and domestic heating systems have a boiling range of 175–360 °C. Any material boiling at a higher temperature is used, if distilled, as (low speed) marine diesel fuel, atomising burner fuel and lube oil base. Undistilled, it is used as residual fuel for use in industry and power generation.*

Practically all the above mentioned petroleum fractions, *i.e.* distillates and residual fuels, can be gasified, and it is proposed to discuss briefly the routes to SNG that one would use and the particular characteristics of the feedstock that will ensure satisfactory gasification performance. Typical properties of the distillate fractions to be reviewed in the following sections are listed in Table 4.7. Similar data for residual fuel oils are given in Tables 4.11 and 4.12.

---

* Appendix B shows a simplified flowsheet of a typical 'integrated' refinery which incorporates distillation, light ends reforming and cracking units contributing to the production of feedstocks for the manufacture of SNG.

TABLE 4.7
Typical Properties of Refinery Products (LPG to Gas Oil)
(Feedstocks for SNG Production)

	Propane	Butane	Light naphtha	Heavy naphtha	Gas oil
Specific gravity 15/15°C	0·510	0·575	0·64–0·72	0·75	0·838
Flash point (PM) °C			Negligible		+65
Viscosity cP, 15°C	0·10	0·15	1·0	3·5	—
cP, 37·8°C	—	—		—	3·3
Ash, % wt	Nil	Nil			0·0002
Sulphur, % wt	0·0015	0·0015	0·006	0·035	0·8
Sediment, mg/100 ml	Nil	Nil	Nil	Nil	0·2
Distillation, °C					
Initial boiling point	−45	−7	35	100	215
50%	—	—	70	135	280
90%	—	—	—	—	320
Final boiling point	−38·3	2·2	150	200	360
Calorific value					
Gross, Btu/lb	21 500	21 200	20 500	20 000	19 525
kcal/kg	11 945	11 779	11 390	11 112	10 850
Net, Btu/lb	20 000	19 800	19 300	18 900	18 310
kcal/kg	11 112	11 000	10 723	10 500	10 173
Carbon, % wt	81·85	82·71	83·7	84·7	86·5
Hydrogen, % wt	18·15	17·29	16·3	15·3	13·5
Aromatics, % vol	Nil	Nil	3–8	8–10	20
Olefins, % vol	5–20[a]	<10[b]	<1	1	2
Saturates, % vol	80–95 (C$_3$/C$_4$)	>90 (C$_3$/C$_4$)	95	90	78

[a] Propene.
[b] Butenes.

4.3.2.1 *Liquefied Petroleum Gas (LPG)*

LPG can be produced either as a result of refining crude oil in standard petroleum processing equipment, or alternatively from gas condensate obtained in the course of the purification of natural gas. LPG consists principally of the C_3/C_4 hydrocarbons—propane/propene and butanes/butenes respectively—together with minor proportions of ethane and pentanes. Impurities content is generally low since purification of the gas is relatively easy. Specifications exist that strictly define the composition and properties of the material which is usually made available in three grades, LPG propane, LPG butane and LPG mixture.

LPG can be used in SNG manufacture in two ways: either the light hydrocarbon feedstock is catalytically reformed in the presence of steam to produce a mixture of methane, hydrogen and carbon dioxide, which is subsequently further treated to produce interchangeable SNG, or alternatively the gas is blended as such into other SNG components, generally in order to increase the calorific value of a gas that is otherwise too lean to meet specifications.

The low molecular weight of LPG components and their simple chemical composition facilitate purification such as removal of sulphur compounds, water vapour, etc. and the conversion equipment is consequently simpler in the case of LPG than it is for the next higher petroleum fraction, naphtha or light distillate. Nevertheless it is rare for SNG plant to be exclusively designed for LPG conversion since availability of LPG is generally limited and furthermore storage of LPG is more expensive than that of naphtha, the gas having to be liquefied for storage by either compression or refrigeration. In consequence, conversion equipment is usually designed to handle either LPG or naphtha. Use of LPG for trimming the heating value of SNG on the other hand is more common and, in fact, the addition of LPG to lean gases up to the limit of acceptable dew points is one of the more economic ways of blending specification gases.

Typical properties of the various LPG grades marketed mainly for other purposes such as automotive fuels, domestic bottle gas, industrial fuels and solvent applications, are shown in Table 4.7. Commercial specifications exist in most industrial countries and have been critically evaluated in a recent publication.[1] They have one weakness as far as SNG feedstock is concerned in that they often do not distinguish between saturated propane and unsaturated propene.

While this may not be of great significance for a number of LPG uses, particularly cooking and heating, it would affect the performance of LPG as an SNG feedstock. Gas industry specifications, to the extent that LPG has been used in the past for the manufacture of lean gases, have therefore limited unsaturated components to between 5 and about 20 % vol. This restriction has applied more particularly to LPG from crude oil refineries where distillate cracking plants can contribute by-product olefinic gases to LPG production. In contrast, LPG which derives from natural gas wells contains minimal quantities of unsaturated hydrocarbons. The other problem that might arise in utilising commercial grades of LPG for SNG manufacture is the presence of stenching compounds, which frequently have to be added to bottle gas for safety reasons. For SNG manufacture it would be necessary either to omit the addition of odorant or, if this was impossible, to remove the odorising compounds, usually organic sulphur compounds, in the first processing step, *e.g.* by desulphurisation.

In the light of the strong demand for LPG as a clean burning industrial fuel, as well as for bottled gas and as a source of automotive energy, it is not considered likely that future SNG plants will be designed for LPG processing. More attention will therefore have to be paid to the heavier feedstocks derived from crude oil.

4.3.2.2 *Naphtha*

Virgin naphtha, also referred to as primary flash distillate or light distillate, is the next heavier fraction after LPG obtained from crude petroleum. The typical properties of the material listed in Table 4.7 demonstrate both advantages and drawbacks in comparison with LPG.

In particular, naphtha has the advantage of a much higher initial boiling point and it can therefore be stored in atmospheric tanks without pressurisation or refrigeration. Its yield from crude oil is also much greater than that of LPG and its availability is therefore somewhat higher. Furthermore, in addition to naphtha from crude oil, materials of very similar composition are obtained from natural gas by condensation and also occur in so-called gas-condensate fields (Section 4.3.2.3).

On the debit side, naphtha as a feedstock for SNG has to compete with alternative outlets such as the manufacture of motor fuel and its use as a petrochemicals feedstock. Before catalytic conversion, naphthas also have to be extensively purified since the sulphur content of straight-run materials exceeds the tolerances of commercial catalysts. Finally, the presence of aromatics and olefins, in addition to saturated hydrocarbons, in some naphthas tends to lower the conversion efficiency of certain SNG routes.

As with LPG, naphtha feedstocks are normally converted into SNG by a low temperature steam reforming process followed by enrichment, *e.g.* by methanation and CO_2 removal. The high nickel content catalyst used in commercial reformers is protected by a desulphurisation system which converts sulphur compounds into H_2S and absorbs the latter on a suitable absorbent such as zinc oxide or Luxmasse. Clearly an excessively high sulphur content will necessitate frequent replacement of the latter. Alternatively, naphthas can be hydrogasified by reacting them under pressure with a hydrogen-rich gas. While the presence of sulphur compounds in the feedstock does not prevent this type of conversion taking place, it can result in excess hydrogen sulphide being present in the hydrogenated product, which then has to be removed before further processing. This, again, results in added cost.

Neither catalytic steam reforming nor hydrogenation is particularly effective for the conversion of aromatics; while a small amount of aromatics gasification seems to occur when steam reforming light feeds, the passage of naphtha through a gas recycle hydrogenator, for instance, leaves benzene nuclei entirely untouched. Gasification feedstocks of high aromaticity therefore result in the formation of considerable amounts of a liquid aromatic by-product.

Finally, olefins and other unsaturates undergo an exothermic reaction with hydrogen in both catalytic reformer hydro-desulphurisation trains and hydrogasifiers, consuming in each instance expensive hydrogen destined for other more essential chemical conversions.

Since refractory sulphur compounds, aromatics and olefins generally all increase in concentration as one moves from lighter to heavier petroleum fractions, an easy means of ensuring naphtha gasification quality in all three respects is to limit its final boiling point. But clearly this militates against the need to maximise naphtha production in order to increase its availability and keep feedstock prices low. Not surprisingly this has produced some conflicts in regard to feedstock specifications for SNG plants.

The principal components of naphthas are hydrocarbons in the range from C_5 to C_9, with a preponderance of C_5 to C_8 in light and C_6 to C_9 in heavy naphtha. However, hydrocarbons of different chemical structure, even if their carbon number is the same, may have different boiling points, and also, owing to interaction effects, naphthas of the same boiling range may well have different hydrocarbon make-ups. This is shown in Table 4.8 which compares two naphthas of the same boiling range but of different crude origin, whose compositions are significantly different.

TABLE 4.8
Hydrocarbon Composition of Two Light Virgin Naphthas Boiling Range C_5–150 °C

Composition, % wt	Naphtha A		Naphtha B	
Paraffins	85·8		80·6	
C_4		7·0		—
C_5		32·0		31·1
C_6		29·4		26·6
C_7		10·8		12·1
C_8		6·6		10·8
Naphthenes	8·9		6·4	
C_5		1·4		1·5
$C_6{}^a$		5·0		3·2
C_6		2·5		1·7
Aromatics	2·1		1·5	
C_6		1·2		0·4
C_7		0·9		1·1
Other	3·1		11·6	
Other C_8		2·5		3·7
C_9 and higher		—		7·4
Not recognised		0·6		0·5

^{a} Gas chromatography peaks for certain C_7 paraffins and C_6 naphthenes overlap; hence the uncertainty.

Both naphthas, being prepared from crude oil by a simple distillation process, are low in aromatics and free of unsaturates. Secondary refining processes on the other hand, which are generally employed to convert straight distillates of low octane number into motor spirit by thermal or catalytic cracking, thermal or catalytic reforming and other routes, increase aromaticity and sometimes unsaturation. For gasification feedstocks they are therefore not only wasteful but positively undesirable, and as a result it has frequently been specified that gasification naphtha must be straight-run or virgin.

Such a qualification, however, would rule out the non-aromatised portion of naphtha that has been steam reformed to produce benzene–toluene aromatics. In fact, these aromatics have often been solvent-extracted and the raffinate is then suitable for blending with virgin naphtha and routing to gas production.

While it is therefore true that the two criteria of restricted boiling range and preparation by simple distillation will generally ensure acceptable gasification performance, it would seem that there are more effective ways of guaranteeing a high gasification efficiency. In particular, the need to feed essentially paraffinic hydrocarbons, with a second preference for naphthenes—the latter when heated can decompose into aromatics and hydrogen—can be satisfied by demanding a minimum content of saturated hydrocarbons. Similarly, a maximum sulphur content should be specified. Finally, setting a maximum percentage of aromatic and olefinic components is distinctly preferable to specifying that the material must be prepared by straight fractionation.

Another aspect of gasification naphtha quality is connected with the various desulphurisation treatments that petroleum products undergo before they reach the gasification system proper. It is common practice to 'sweeten' naphthas before dispatch from a refinery or natural gasoline plant. In this process strongly odorous organic sulphur compounds such as mercaptans are converted either into H_2S which is removed (hydrofining), or into sulphides and disulphides (oxidative sweetening) which may be solvent extracted. Neither process is particularly effective as far as removal of aromatic sulphur compounds is concerned, and these so-called 'non-reactive' forms of sulphur are also less likely to be converted by the purification systems of the SNG plant and may therefore penetrate into the steam reforming section and deactivate its catalyst. Relatively little is known about the occurrence of non-reactive sulphur compounds in naphthas of different origin. But occasionally, and as a precaution, a maximum content may be specified in feedstock purchase contracts.

Summing up, therefore, the following naphtha characteristics are considered relevant as far as SNG manufacture is concerned and the following limiting values have been set up in attempts to standardise feedstocks and gasification plant design.

Paraffin content, % vol	80 min
Aromatics content, % vol	5 or 10 max
Unsaturates content, % vol	2 max

Total sulphur content, ppm   20 or 100 or 500
Unreactive sulphur, ppm   20 max
Distillation (Final B.P., °C)   150 or 185 or 200
The feedstock to be straight run

Of these the first four seem entirely rational, the fifth, unreactive sulphur, is probably justified, but the last two, distillation range and insistence on one particular route of manufacture, are of only limited significance.

4.3.2.3 Natural Gas Condensate

Since naphtha is a preferred feedstock for SNG manufacture but is in relatively short supply, much interest centres on the availability and suitability of natural gas liquids.

Natural gas liquids consist of hydrocarbons boiling in the C_3–C_{10} range and can be obtained from both 'associated' and 'dry' gases. After preliminary crude oil/gas separation, the latter is cooled and scrubbed and the condensed propane, butane and naphtha fractions are drawn off. Similarly 'dry' gases, although they contain less condensable material, must have their LPG content reduced or eliminated before they are dispatched by pipeline. If natural gas is re-injected for secondary oil recovery, it is also necessary to separate condensable liquids in order to compress the purified gas.

Current statistics show the total world recovery of liquids to amount to about 100 million tons with the USA being the leading producer accounting for some 64 million and Canada, the Middle East countries and the USSR coming almost equally in second place (7–10 million tons). Table 4.9 gives a breakdown of liquids production into LPG and naphtha 'condensate' portions.

However, this picture may change as additional recovery schemes planned particularly for Middle East countries, Australia and the North Sea get underway.

As regards quality, condensate resembles naphtha in appearance and gravity, and some condensates resemble light virgin naphtha in having an end boiling point below 180 °C, low sulphur content (0·01 % wt) and little or no aromatic or olefinic hydrocarbon content. Others, as recovered from the gas fields, have a sizable tail boiling above 180 °C, the maximum end boiling point often specified by SNG plant designers and contractors. The presence of such high boiling ends may lead to increased coking on the reforming catalyst as well as

TABLE 4.9
World Production (1 000 tons) of Natural Gas Liquids in 1973
Main Producers

	Propane	*Butane*	*Condensate*	*Total*
USA	17 572	8 506	37 640	63 718
Canada	2 830	2 127	7 180	12 137
USSR	—	—	—	7 470
Saudi Arabia		2 160	1 120	3 280
Venezuela		1 960	1 200	3 160
Mexico		1 880	490	2 370
Kuwait	700	717	670	2 087
Algeria	—	—	1 420	1 420
Libya	43	211	1 140	1 394
Iran	405	384	520	1 309

Source: Energy Pipelines Systems, **1**(10), 36, 1974.

lowering gasification efficiency. Often though, natural gas condensate is reprocessed at oil refineries, in which case it will be cut appropriately along with the straight-run naphtha fractions.

Aromatics content is another variable property, some condensates showing less than 5 % vol, like virgin naphthas, others showing up to 20 % vol or more. Whilst not rendering condensates unsuitable for steam reforming processes, aromatics and end boiling point are two characteristics that should be monitored when evaluating SNG manufacturing capability.

Table 4.10 provides some inspection data on natural gas condensates.

TABLE 4.10
Range of Properties of Natural Gas Condensate

Specific gravity, 15/4 °C	0·68–0·78
Sulphur, % wt	0·015
Distillation, °C	
I.B.P.	35
50 %	160
95 %	275
F.B.P.	285
% vol boiling below 182 °C (carbon range C_5–C_{10})	65–95
Composition, % vol	
Paraffins	65–75
Naphthenes	20–30
Aromatics	3–20
Olefins	< 1

4.3.2.4 *Gas Oils*

Gas oils and distillate fuel oils are distilled from crude oil at temperatures somewhat higher than those for naphtha, their yield and exact boiling range depending on type of crude oil processed, demand for lighter products such as kerosine and naphtha, and gravity and mix of heavier, residual fuel oils, which all affect the cut points of the fraction. Since automotive diesel fuel, heating oils and some petrochemical feedstocks are based on the same products, sales of gasification feedstocks will have to compete with these alternative outlets.

Sulphur content, boiling range and sometimes unsaturates content of gas oils militate against their use as steam reformer feedstocks and it is more common to gasify middle distillates by hydrogenation. However, aromatics content and therefore yield of by-product aromatic liquids can be considerable and consequently gasification efficiency tends to be low.

Typical gas oil characteristics are listed in Table 4.7. It should be noted, however, that considerable variations in practically all the figures listed will frequently occur.

4.3.2.5 *Fuel Oils*

These oils are composed of the residuum remaining after distillation of the lighter ends of crude oil, cut back to an appropriate viscosity with straight-run middle distillate or cracked gas oil. Three grades of commercial fuel oils that differ mainly in viscosity are marketed in most countries under the names of light, medium and heavy. Viscosities are generally adjusted to meet national specifications and/or the internationally recognised ASTM standard D396, reproduced as Table 4.11, in which grades 1 to 4 refer to distillates and grades 5 to 6 to residual fuel oils.

The main difference between the three viscosity grades is one of handling. Light fuel oils of a maximum viscosity of 65 cSt at 38 °C do not normally have to be preheated in storage or transfer lines, except in very cold climates. Medium fuel oil of no more than 162 cSt at 38 °C requires preheat in storage, particularly during the winter, since wax would settle out or the material would set at any temperature lower than the pour point. The oil must also be heated to 66 °C for efficient atomisation in an injection nozzle. The heaviest grade, also known as bunker fuel C, can have a viscosity of up to 638 cSt at 50 °C and

TABLE 4.11
ASTM D396
Detailed Requirements for Fuel Oils[a]

Grade of fuel oil	Flash point, deg. F (deg C)	Pour point, deg. F (deg. C)	Water and sediment volume (%)	Carbon residue on 10% bottoms (%)	Ash (weight %)	Distillation temperatures deg F (deg C)		
	Min.	Max.	Max.	Max.	Max.	10% point Max.	90% point Min.	90% point Max.
No. 1 A distillate oil intended for vaporising pot-type burners and other burners requiring this grade of fuel	100 or legal (38)	0[c]	trace	0·15	—	420 (215)	—	550 (288)
No. 2 A distillate oil for general purpose domestic heating for use in burners not requiring No. 1 fuel oil.	100 or legal (38)	20[c] (−7)	0·05	0·35	—	d	540[c] (282)	640 (338)
No. 4 Preheating not usually required for handling or burning	130 or legal (55)	20[c] (−7)	0·50	—	0·10	—	—	—
No. 5 (Light) Preheating may be required depending on climate and equipment	130 or legal (55)	—	1·00	—	0·10	—	—	—
No. 5 (Heavy) Preheating may be required for burning and, in cold climates, may be required for handling	130 or legal (55)	—	1·00	—	0·10	—	—	—
No. 6 Preheating required for burning and handling	150 (65)	h	2·00[f]	—	—	—	—	—

[a] It is the intent of these classifications that failure to meet any requirement of a given grade does not automatically place an oil in the next lower grade unless in fact it meets all requirements of the lower grade.
[b] In countries outside the United States other sulphur limits may apply.
[c] Lower or higher pour points may be specified whenever required by conditions of storage or use. When pour point less than 0 °F is specified, the minimum viscosity for Grade No. 2 shall be 1·8 cSt (32·0 s Saybolt Universal) and the minimum 90% point shall be waived.
[d] The 10% distillation temperature point may be specified at 440 °F (226 °C) maximum for use in other than atomizing burners.
[e] Viscosity values in parentheses are for information only and not necessarily limiting.
[f] The amount of water by distillation plus the sediment by extraction shall not exceed 2·00%. The amount of sediment by extraction shall not exceed 0·50%. A deduction in quantity shall be made for all water and sediment in excess of 1·0%.
Reprinted from copyright protected material published by ASTM.

requires storage temperatures of about 40 °C and an injection temperature of at least 120 °C for acceptable atomisation.

Table 4.12 lists typical properties of commercial fuel oils, and again it should be noted that characteristics of these products, in the light of their wide range of origins and methods of manufacture, will show fairly extensive variations.

Fuel oils can be gasified by partial oxidation with oxygen or air and by fluid bed hydrogenation, and most grades are acceptable feedstocks for all the main process routes. Impurities such as sulphur or nitrogen in the feedstock do not interfere with gasification as such but impose a heavier gas cleaning load. High viscosities and pour points are also acceptable but lead to the additional expense for

FEEDSTOCKS FOR SNG MANUFACTURE 77

TABLE 4.11
ASTM D396
Detailed Requirements for Fuel Oils^a—continued

Saybolt viscosity, s^e				Kinematic viscosity, cSt^e				Gravity (deg. API)	Copper strip corrosion	Sulphur (%)
Universal at 100°F (38°C)		Furol at 122°F (50°C)		At 100°F (38°C)		At 122°F (50°C)				
Min.	Max.	Min.	Max.	Min.	Max.	Min.	Max.	Min.	Max.	Max.
—	—	—	—	1·4	2·2	—	—	35	No. 3	0·5 or legal
32·6)	(37·93)	—	—	2·0^c	3·6	—	—	30	—	0·5^b or legal
45	125	—	—	(5·8)	(26·4)^c	—	—	—	—	legal
150	300	—	—	(32)	(65)^g	—	—	—	—	legal
350	750	(23)	(40)	(75)	(162)^g	(42)	(81)	—	—	legal
(900)	(9 000)	45	300	—	—	(92)	(638)^e	—	—	legal

Where low sulphur fuel oil is required, fuel oil falling in the viscosity range of a lower numbered grade down to and including No. 4 may be supplied by agreement between purchaser and supplier. The viscosity range of the initial shipment shall be identified and advance notice shall be required when changing from one viscosity range to another. This notice shall be in sufficient time to permit the user to make the necessary adjustments.

Where low sulphur fuel oil is required, Grade 6 fuel oil will be classified as low pour (60°F max) or high pour (no max). Low pour fuel oil should be used unless all tanks and lines are heated.

urtesy: American Society for Testing and Materials.

feedstock preheat. Aromatics and complex hydrocarbons do not greatly affect performance in partial oxidation, although they may marginally increase carbon black forming tendency, but are not desirable in fluid bed hydrogenation since they are converted into liquid aromatic by-products instead of gas. A parameter of some significance is Conradson Carbon Number, a measure of the tendency to form carbon deposits by pyrolysis. Clearly, in order to facilitate preheating ahead of feedstock injection and also to allow alternative gasification reactions to take place in preference to thermal cracking a low carbon-forming tendency will be at a premium. Finally, metal and ash content can result in deposits on the walls of process vessels and interfere with process operability; a low ash content in fuel oils used for gasification is therefore also desirable.

TABLE 4.12
Typical Properties of Commercial Fuel Oils

	Light	Medium	Heavy
Viscosity, Saybolt Universal, Sec. 37.8°C	200–350	750–1 500	2 500–6 000
kinematic, cSt, 82.2°C	10.0	29.0	58
Specific gravity, 15/4°C	0.93	0.945	0.960
Pour point, °C	fluid at 18	−4 to +7	10 to 24
Flash point, C.P.M.,[a] °C min	65.6	65.6	65.6
Water, % vol	0.1	0.1	0.1
Sediment, % wt	0.03	0.04	0.04
Ash, % wt	0.04	0.04	0.04
Sulphur, % wt	2.0–2.5	2.0–3.0	1.0–3.5[b]
Calorific value:			
Gross, Btu/lb	18 670	18 570	18 450
Btu/IG	173 680	175 490	171 000
Btu/USG	145 000	150 000	142 500
Net, Btu/lb	17 600	17 520	17 380
Btu/IG	163 680	165 560	166 800
Btu/USG	136 400	137 970	139 000
Carbon, % wt	85.9	85.9	85.6
Hydrogen, % wt	11.6	11.6	11.4
Vanadium, ppm	35	45	60
Asphaltenes, % wt	1.9	2.2	2.5
Conradson carbon, % wt	8.1	8.8	10.1

[a] Closed Pensky–Martens.
[b] Special low sulphur grade, not freely available.

4.4 CONCLUSIONS

Almost any hydrocarbon fuel from LPG, the lightest, to bituminous coal, the heaviest, can be gasified by appropriate conversion processes and ultimately processed to make acceptable SNG. However, owing to increasing molecular complexity, lower hydrogen/carbon ratio and higher degree of contamination, as one goes up the scale of average molecular weights, the heaviest materials will be more difficult and more expensive to gasify than products such as LPG or naphtha. Thus light hydrocarbon products will be at a premium as gasification feedstocks and plants based on naphtha, LPG or condensate will be built wherever SNG is required and a commercially viable supply of light feedstock can be made available. However, competitive demand for these light hydrocarbon products is strong and their prices are high. Very large volumes can rarely be made available. Heavier materials, especially crude oil, have therefore been considered as alternative feedstocks. But in the light of recent world-wide shortages and price increases for all liquid hydrocarbons, coal is by now probably the only gasification feedstock that is available in sufficient volumes and at prices that are unlikely to be forced upwards excessively by competing outlets.

REFERENCES

1. Williams, A. F. and Lom, W. L. (1974). *Liquefied Petroleum Gases*, Wiley, Chichester, New York.
2. Moore, J. F. (1974). A look at SNG feedstocks, *Energy Pipelines and Systems*, January, 39.
3. Roeger III, A. (1973). Liquid feed offers quick way to SNG, *Oil Gas J.*, June, 110.
4. Hobson, G. D. and Pohl, W. (Eds.) (1973). *Modern Petroleum Technology*, 4th Ed., Applied Science Publishers, Barking.

Chapter 5

Gasification Systems

5.1 INTRODUCTION

The aim of all gasification processes is to convert a fossil fuel of high molecular weight, low volatility, high carbon to hydrogen ratio and, frequently, high impurity content, into a low molecular weight, gaseous, low C/H ratio and ultimately clean fuel suitable for pipeline distribution. In the case of SNG processes, furthermore, it is intended to reproduce in the manufactured gas as closely as possible all the properties and specially the combustion characteristics of natural gas.

In the present chapter it is proposed to discuss the chemical reactions involved in these conversions, and to emphasise the basic molecular changes that the hydrocarbon molecules undergo in the course of their progression from heavy liquid or solid to a gas. Little or no attention will be paid at this stage to the various impurities and minor components of the feedstock. Similarly the discussion of industrial process routes to SNG will be relegated to subsequent chapters, our main concern here being the chemical and physical principles involved in such conversions.

5.2 BASIC PRINCIPLES[2,5,6]

In order to convert a heavy hydrocarbon of a carbon to hydrogen atomic ratio of 1:2 or higher into methane, which has a C/H ratio of 1:4, one can clearly either increase the hydrogen or decrease the carbon content of the molecule.

To increase the hydrogen content of the system one can introduce hydrogen in gaseous form and, provided conditions are right for the two components to react, a gasification reaction which is sometimes termed 'hydrogenolysis' will take place. Alternatively, where elementary hydrogen is not available, one can generate hydrogen within the system from steam, again given the right operating conditions, and it will react with the hydrocarbon feedstock much as external hydrogen would. In this case, lighter hydrocarbons including methane, will be formed by a process which is known as 'hydrolysis'.

To reduce the carbon content of the system it is treated in such a way as to form a product—carbon dioxide, coke or char—that can be removed from the system. To form carbon dioxide, oxygen must be added and the corresponding gasification process is described as 'oxygenolysis'. The carbon dioxide is usually removed by scrubbing the gas with amine or hot potassium carbonate solution.

Char is normally formed by heating the hydrocarbon to cracking temperature, and the corresponding gasification step has therefore been referred to as 'pyrolysis'.

Basically all gasification processes can be reduced in principle to one or more of the above reactions and it is, as a rule, possible to classify them as conversions involving mainly hydrogenolysis, hydrolysis, oxygenolysis or pyrolysis or a combination of these reactions.

The overall chemical reaction in the four possible gasification systems can be expressed as:

$$\text{hydrogenolysis—} C_n H_m + \frac{4n - m}{2} H_2 \rightarrow n\,CH_4$$

$$\text{hydrolysis—} C_n H_m + x\,H_2O \rightarrow y\,CH_4 + z\,CO_2 + w\,H_2$$

$$\text{oxygenolysis—} C_n H_m + \frac{4n - m}{4} O_2 \rightarrow \frac{m}{4} CH_4 + \frac{4n - m}{4} CO_2$$

$$\text{pyrolysis—} C_n H_m \rightarrow \frac{m}{4} CH_4 + \frac{4n - m}{4} C$$

However, this is only true in an overall sense and a large number of intermediates are in fact formed in the course of this overall conversion. The chemical equilibria that control the gasification yields of methane, CO_2, carbon, etc., are the reactions that are listed in Table 5.1 together with their heat of reaction, ΔH, the gaseous volume change involved, ΔV, and the form of the equilibrium constant K_p.

The last column in the table expresses the equilibrium constant K_p, the values of which at different temperatures will be quoted later. The volume change listed in column 3 indicates whether the effect of a

TABLE 5.1
Equilibrium in Gasification Reactions

No.	Reaction	Heat of reaction ΔH (kcal/kg mole at 298·16 K (25 °C))	Volume change (ΔV)	Equilibrium constant (K_p)
1	$C + O_2 \rightarrow CO_2$	$-94\,056$	1 to 1	$\dfrac{(CO_2)}{(O_2)}$
2	$C + CO_2 \rightarrow 2CO$	$+41\,203$	1 to 2	$\dfrac{(CO)^2}{(CO_2)}$
3	$C + H_2O \rightarrow H_2 + CO$	$+31\,356$	1 to 2	$\dfrac{(H_2)(CO)}{(H_2O)}$
4	$CO + H_2O \rightarrow H_2 + CO_2$	-9847	2 to 2	$\dfrac{(H_2)(CO_2)}{(CO)(H_2O)}$
5	$CO + 3H_2 \rightarrow CH_4 + H_2O$	$-49\,243$	4 to 2	$\dfrac{(CH_4)(H_2O)}{(CO)(H_2)^3}$
6	$C + 2H_2 \rightarrow CH_4$	$-17\,889$	2 to 1	$\dfrac{(CH_4)}{(H_2)^2}$

7	$CO_2 + 4H_2 \rightarrow CH_4 + 2H_2O$	$- 39\,410$	5 to 3	$\dfrac{(CH_4)(H_2O)^2}{(CO_2)(H_2)^4}$
8	$C + 2H_2O \rightarrow 2H_2 + CO_2$	$+ 21\,510$	2 to 3	$\dfrac{(H_2)^2(CO_2)}{(H_2O)^2}$
9	$CH_4 + CO_2 \rightarrow 2CO + 2H_2$	$+ 59\,093$	2 to 4	$\dfrac{(CO)^2(H_2)^2}{(CH_4)(CO_2)}$
10	$2C + H_2 \rightarrow C_2H_2$	$+ 54\,163$	1 to 1	$\dfrac{(C_2H_2)}{(H_2)}$
11	$2CH_4 \rightarrow C_2H_2 + 3H_2$	$+ 89\,918$	2 to 4	$\dfrac{(C_2H_2)(H_2)^3}{(CH_4)^2}$
12	$2CH_4 \rightarrow C_2H_4 + 2H_2$	$+ 48\,246$	2 to 3	$\dfrac{(C_2H_4)(H_2)^2}{(CH_4)^2}$
13	$C_2H_6 \rightleftharpoons C_2H_4 + H_2$	$+ 32\,710$	1 to 2	$\dfrac{(C_2H_4)(H_2)}{(C_2H_6)}$
14	$CH_4 + H_2O \rightarrow CO + 3H_2$	$+ 49\,242$	2 to 4	$\dfrac{(CO)(H_2)^3}{(CH_4)(H_2O)}$

Source: A.G.A., Gas Engineer's Handbook. (Courtesy: Industrial Press, N.Y.)

pressure increase would be to drive the equilibrium to the right (volume reduction) or to the left (volume increase). The heat of reaction at 25 °C is a measure of the degree of endothermicity or exothermicity of the reaction, *i.e.* it indicates whether heat must be supplied (ΔH positive) or must be withdrawn (ΔH negative) in order to drive the reaction towards the right.

The significance of the 14 reactions listed is that they invariably occur in the course of the overall conversion from liquid or solid feedstock to gas. On the other hand, a gasification or SNG process cannot be based simply on one or even a few of these conversions, but at some stage practically all 14 equilibria coexist.

There is very little doubt for example that methane cannot be formed directly from higher hydrocarbon molecules and steam and that carbon oxide intermediates are invariably formed. In other words, steam reacts with the hydrocarbon to produce carbon monoxide and hydrogen and the latter react in accordance with the methanation reaction (5). Additional hydrogen to assist the formation of more methane is normally obtained by reaction (4), the carbon monoxide shift.

Similarly the destructive hydrogenation of hydrocarbon molecules may well proceed by way of a preliminary cracking stage to olefins or even to carbon which, in turn, is hydrogenated to methane according to reaction (6). Oxidative attack on hydrocarbons is generally preceded by thermal cracking (13) and the oxidation of the carbon atom (1) may be controlling.

An important consideration is the enthalpy change ΔH accompanying the different steps. The oxidation of carbon and of hydrocarbons is strongly exothermic but the steam reforming of hydrocarbons (14), the water gas reaction (carbon plus steam, 3) and the shift reaction (4) are endothermic not only at 25 °C but at temperatures up to the reaction temperature and beyond. The hydrogenation of carbon to paraffins is exothermic but the formation of acetylene and ethylene either from the component elements or by cracking of hydrocarbons is endothermic. Finally, methanation of carbon oxides, reaction (5), is strongly exothermic.

In order to produce a gas from higher molecular weight materials in the most efficient and economic fashion it is essential to balance these effects, *i.e.* to minimise the thermal energy that has to be supplied from outside. One of the features of SNG manufacture is the final methanation of residual carbon oxides. Utilisation of the excess heat generated in this step in some of the prior conversions will clearly improve its economics. Close heat integration of the different steps, if for other reasons they cannot be combined, will therefore be desirable.

The justification of stepwise processing is generally the need for different process conditions in the various stages of gasification. It may for example be desirable to increase pressure at a certain stage or to change the temperature. The effect of pressure on the reactions considered in Table 5.1 has already been discussed. The effect of a temperature change at constant pressure needs some further consideration.

Table 5.2 lists values of K_p, the equilibrium constant of each reaction, at a number of different temperatures. Where K_p tends to increase as the temperature is raised a temperature rise will shift the reaction equilibrium to the right (e.g. 2, 3, 9); in the case of a lowering of K_p with increase in temperature, on the other hand (e.g. in 1, 4 and 6), a temperature decrease will tend to drive the reactions towards completion.

While the value of K_p at a given temperature establishes the composition of the final gas mixture once chemical equilibrium between the different species has been established, it gives no indication whether this will occur after a short or long period. This depends on the velocity constants of both forward and reverse reactions, which in turn are a function of the reactivity of the different species. The latter can be affected very profoundly by the presence or absence of certain types of surface, the catalyst, and for many gasification reactions the presence of a catalyst is essential in order to reach a state of equilibrium within a reasonable time. Particularly at low temperatures, when the reactivity of the different reactants is low, the presence of a catalyst becomes essential if chemical equilibrium is to be attained.

However, the use of catalysts, particularly of the more active type, carries with it certain problems. Since heterogeneous catalysts, that is materials on the surface of which the reactions between gas molecules take place, act owing to the surface forces exerted by certain active materials or centres, they tend to attract molecules other than those that it is intended to react. Specifically, polar impurities in the feedstock, such as sulphur, nitrogen, oxygen and metal compounds will deposit on the catalyst, block the active centres and eventually poison the catalyst. Some catalysts can be regenerated by removing the deposits with oxygen, steam or heat, but others are permanently

SUBSTITUTE NATURAL GAS

TABLE 5.2
Value of Equilibrium Constant K_p in Gasification Reactions

No.	Equilibrium constant (K_p)	Value of K_p		
		at 800 K	1000 K	1200 K
1	$\dfrac{(CO_2)}{(O_2)}$	$6 \cdot 709 \times 10^{25}$	$4 \cdot 751 \times 10^{20}$	$1 \cdot 738 \times 10^{17}$
2	$\dfrac{(CO)^2}{(CO_2)}$	$1 \cdot 098 \times 10^{-2}$	$1 \cdot 900$	$57 \cdot 09$
3	$\dfrac{(H_2)(CO)}{(H_2O)}$	$4 \cdot 399 \times 10^{-2}$	$2 \cdot 609$	$39 \cdot 77$
4	$\dfrac{(H_2)(CO_2)}{(CO)(H_2O)}$	$4 \cdot 038$	$1 \cdot 374$	$0 \cdot 6966$
5	$\dfrac{(CH_4)(H_2O)}{(CO)(H_2)^3}$	$8 \cdot 821 \times 10^{-2}$	$4 \cdot 383$	$59 \cdot 585$
6	$\dfrac{(CH_4)}{(H_2)^2}$	$1 \cdot 411$	$9 \cdot 829 \times 10^{-2}$	$1 \cdot 608 \times 10^{-2}$
7	$\dfrac{(CH_4)(H_2O)^2}{(CO_2)(H_2)^4}$	$2 \cdot 185 \times 10^{-2}$	$3 \cdot 190$	$85 \cdot 537$
8	$\dfrac{(H_2)^2(CO_2)}{(H_2O)^2}$	$0 \cdot 1777$	$3 \cdot 608$	$28 \cdot 01$
9	$\dfrac{(CO)^2(H_2)^2}{(CH_4)(CO_2)}$	$7 \cdot 722 \times 10^{-3}$	$19 \cdot 32$	$3 \cdot 548 \times 10^3$
10	$\dfrac{(C_2H_2)}{(H_2)}$	$1 \cdot 602 \times 10^{-12}$	$1 \cdot 337 \times 10^{-9}$	$1 \cdot 156 \times 10^{-7}$
11	$\dfrac{(C_2H_2)(H_2)^3}{(CH_4)^2}$	$8 \cdot 017 \times 10^{-13}$	$1 \cdot 384 \times 10^{-7}$	$4 \cdot 474 \times 10^{-4}$
12	$\dfrac{(C_2H_4)(H_2)^2}{(CH_4)^2}$	$1 \cdot 021 \times 10^{-7}$	$6 \cdot 939 \times 10^{-5}$	$5 \cdot 540 \times 10^{-3}$
13	$\dfrac{(C_2H_4)(H_2)}{(C_2H_6)}$	$4 \cdot 557 \times 10^{-3}$	$0 \cdot 3443$	$6 \cdot 224$
14	$\dfrac{(CO)(H_2)^3}{(CH_4)(H_2O)}$	$3 \cdot 120 \times 10^{-2}$	$26 \cdot 56$	$2 \cdot 473 \times 10^3$

Source: A.G.A., Gas Engineer's Handbook. (Courtesy: Industrial Press, N.Y.)

de-activated either by the deposits or by the process used for their removal.

It can therefore be stated quite generally that wherever a gasification process is based on the catalytic conversion of the feedstock it is essential that the latter be purified before it makes contact with the catalyst. Particularly, all sulphur compounds should be removed by hydrotreating and absorption of the hydrogen sulphide formed. This becomes even more important if the catalyst is sensitive to over-heating or steam treating or if the sulphur compounds are so firmly attached that simple catalyst regeneration becomes out of the question.

5.3   STEAM REFORMING REACTIONS (HYDROLYSIS)

The term steam reforming implies the conversion of a hydrocarbon by means of steam and heat into gaseous products, generally in the first instance carbon monoxide and hydrogen. The equation

$$(CH_2)_x + xH_2O \rightarrow xCO + 2xH_2 \qquad (1)$$

gives an overall impression of the reaction of a hydrocarbon in the naphtha range, say C_5 to C_8, with steam at a temperature of about 800 °C when few other side or consecutive reactions occur, or when the above steam reforming reaction can at least be made to predominate over competing reactions by the employment of suitable catalysts. It will be appreciated that an alternative conversion

$$(CH_2)_x + 2xH_2O \rightarrow xCO_2 + 3xH_2 \qquad (2)$$

would be imaginable. However, the higher temperature and the consequent instability of CO_2 v CO will tend to reinforce conversion (1) over (2). Similarly the formation of methane by reaction

$$(CH_2)_x + xH_2O \rightarrow \frac{x}{2}CH_4 + \frac{x}{2} CO_2 + xH_2 \qquad (3)$$

will not occur unless the temperature is well below 600 °C and the methanation reaction equilibrium, therefore, on the right-hand side.

Simple cracking of the hydrocarbon into carbon, or hydrocarbon fragments, and hydrogen will of course also take place at temperatures of 800 °C, particularly if the hydrocarbon feedstock is inherently unstable, as are some of the higher boiling materials. But provided the reforming activity of the catalyst is sufficient to keep carbon deposition at a low level, the carbon regasification reactions (with steam or CO_2) in Tables 5.1 and 5.2 will keep the catalyst clean and active for further high temperature reforming. The presence in the feedstock of impurities that deactivate the catalyst or the rapid formation of carbon from an unstable or heavy feed, will interfere with the continuity of this type of gasification.

The alternative to high temperature steam reforming, reaction (3), will take place at temperatures between 450 and 500° and, in order to initiate a conversion reaction at such a low temperature, extremely active catalysts are required. Such catalysts tend to be deactivated not only by impurities in the feedstock but also by excess temperature or even too high a steam concentration, since their functioning often depends on a particular crystalline structure. Very careful process control is essential under these circumstances.

It is also clear that feedstock purification down to very low contents of contaminants is essential. Sulphur contents in the feedstock of a maximum of about 2 ppm for high temperature reforming have to be reduced to about 0·2 ppm for low temperature reforming. Similarly, feedstock stability becomes even more important, and high boiling point feedstocks are accordingly difficult to process.

Pressurisation has only little effect on the high temperature reforming equilibrium, which it would tend to push towards the left; however, it is normal to gasify under pressure in order to reduce the dimensions of the equipment and also to have the gasified product available under pressure.

The low temperature reforming equilibrium, on the other hand, is influenced favourably by pressurisation, and particularly methanation, sometimes carried out as a separate step, requires a high pressure. All low temperature reforming processes therefore operate at pressures of 20 bar or higher.

Another difference between the two reactions is their enthalpy change; high temperature reforming is endothermic whereas low temperature conversion is exothermic. As a result, the configuration of a high temperature reformer is quite different; the reaction usually takes place in a series of externally fired tubular reactors. Low temperature reforming on the other hand does not require external heating, and gasification therefore usually takes place in an adiabatic reactor, generally an insulated drum packed with catalyst.[4]

5.4   PARTIAL OXIDATION

The gasification of hydrocarbons by partial oxidation is based on the reaction

$$(CH_2)_x + \frac{x}{2}O_2 \rightarrow xCO + xH_2 \qquad (4)$$

However, since oxygen as an oxidising agent is considerably more expensive than air, and the latter, apart from various trace components, consists of 21 % vol oxygen and 79 % vol nitrogen, the more usual conversion can be written as

$$(CH_2)_x + \frac{x}{2}O_2 + 1 \cdot 88x\,N_2 \rightarrow x\,CO + x\,H_2 + 1 \cdot 88\,xN_2 \qquad (5)$$

Consequently a gas produced by partial oxidation of a hydrocarbon with air consists of about 50 % vol inert nitrogen.

The oxidation reaction is, of course, highly exothermic and since further processing of the gas, *e.g.* by methanation, is also exothermic there is no efficient way in which the excess thermal energy produced by partial oxidation can be utilised. The only solution to the dilemma is to combine partial oxidation with some other form of gasification, *e.g.* steam reforming, that is endothermic. This is in fact done commercially and the overall reaction taking place in a partial oxidation gasifier will be of the type

$$(CH_2)_x + \frac{x}{4}O_2 + \frac{x}{2}H_2O \rightarrow xCO + 1\tfrac{1}{2}xH_2 \qquad (6)$$

where the endothermicity of steam reforming is balanced by the exothermicity of partial oxidation.

While it is sometimes desirable to establish oxidation equilibrium with the aid of a catalyst it is more usual to carry out oxygen or air gasification under non-catalytic conditions. Operating temperatures must be high to ensure oxidation in preference to thermal cracking, and pressure, while not essential, helps to keep the equipment within manageable size.

Carbon formation by thermal cracking is not inherently undesirable since steam will react with deposited carbon in preference to fresh feedstock; however, sometimes carbon formed in partial oxidation reactions is of the relatively unreactive carbon black type. If carbon black is formed it usually leaves the reactor with the gas and has to be separated from the latter and recycled. Carbon formation tends to be more pronounced in the gasification of heavy liquid feedstocks, materials that are often gasified in partial oxidation plants because of their high content of sulphur and other contaminants. These do not interfere with the conversion process since there are no catalysts to deactivate or clog in the oxidation stage. Owing to their long carbon to carbon chains and complex molecules, heavy liquid feedstocks tend to break down to simpler molecules which frequently polymerise to solid carbon.

The other main advantage of pressure gasification with air or oxygen is the availability of the product gas under pressure, particularly for gas turbine use or for long-distance distribution. Since industrial oxygen is usually available under pressure and the pressurisation of a liquid or solid feedstock requires little energy, it is normal to operate partial oxidation processes at up to 80 atm, the maximum value depending on further process steps to be used. This high pressure has the added advantage in SNG plants of favouring methane formation from carbon oxides and hydrogen (refer reactions 5 and 7 in Table 5.1).

5.5 HYDROGENATION REACTIONS

The gasification of hydrocarbons by means of hydrogen differs in certain respects from oxidation or steam reforming. In these latter cases, both oxygen and water vapour are freely available and the reaction that occurs is between a feedstock and a reagent requiring little or no prior processing. Hydrogen, on the other hand, must be prepared, normally, from a hydrocarbon, and only the finished product acts as a gasifying reagent.

This is significant if one considers, for example, the enthalpy changes that occur during hydrogasification. Generally the reaction is exothermic but considerable heat energy must be provided to prepare the hydrogen.

Overall hydrogasification can be expressed as

$$(CH_2)_x + xH_2 \rightarrow xCH_4 \tag{7}$$

and from what has been said it will be clear that reaction conditions are characterised by high pressure and as low as practicable a temperature.

In order to utilise the exothermic heat of the hydrocracking reaction it is just possible to link hydrogen production with the hydrogasification step by close heat exchange. However, the problem is complicated by the relatively high initiation temperature for the non-catalytic hydrogenation reaction and thermal integration is therefore practically confined to catalytic hydrogasification. An interesting process balance is used in the second or hydrogasification step of certain low temperature reforming processes, such as the British Gas Corporation's CRG process and the Lurgi Gasynthan process, which operate at a temperature where the reforming step is slightly endothermic, with simultaneous catalytic hydrogenation and methanation reactions balancing the thermal deficit.

The pressure hydrogasification of paraffins and to some extent of naphtha does not appear to require the presence of a catalyst, although it is possible that the surface of stainless steel vessels has some effect on the reactions that take place. It has been shown for instance that on very clean steel surfaces thermal cracking with massive carbon deposition predominates and that hydrogasification occurs only in the presence of small concentrations of sulphur compounds—a negative catalytic effect one might say.

Aromatics present in hydrogasification feedstocks are not converted during the non-catalytic hydrogenation but remain unchanged in the exit gas from which they have to be separated. They can be recycled as process fuel or drawn off as a by-product. This applies to both ring compounds and rings with side chains, the paraffinic part of which appears to be split off and hydrogenated.

In the case of feedstocks heavier than gas oil some carbon is always formed during hydrocracking and provision must be made for its withdrawal. This is done by providing a fluidised bed of char particles on which the additional carbon is allowed to deposit and part of which is continually withdrawn.

5.6 PYROLYTIC CRACKING REACTIONS

While in theory it is possible to produce methane and carbon from hydrocarbons by pyrolysis, there are no commercial gasification processes based on this route. Methane yields would be low and carbon would, as a rule, be formed as a difficult-to-handle carbon black or soot. The character of cracking reactions is therefore one of undesirable side reactions whose extent must be limited, rather than of a route to substitute natural gas.

Nevertheless, methane and other light hydrocarbons are formed by pyrolysis in the course of crude oil refining by atmospheric fractionation, when the uncondensed overhead gas contains hydrogen, methane and ethane. Further volumes of gas result from catalytic cracking, the conversion of gas oil and other middle distillates into gasoline, a refining process that produces substantial quantities of by-product gas (hydrogen, methane, ethane and ethylene), which is withdrawn as a non-condensable overhead stream.

Other refinery processes, *e.g.* coking, steam cracking, visbreaking, hydrocracking, catalytic reforming, while conducted to minimise gas production, still result in some methane formation, owing in most instances to local overheating, insufficient mixing or unsatisfactory process control. The exception is, of course, steam cracking, which converts naphtha or gas oil into ethylene by thermal cracking. Clearly under those circumstances, even if ethylene production is maximised, some methane must be formed.[3]

5.7 METHANATION

An important processing step in the manufacture of SNG is the conversion of carbon oxides into methane by reaction with hydrogen over a catalyst (refer also elaboration in Chapter 10). The reactions

$$CO + 3H_2 \rightarrow CH_4 + H_2O \qquad (8)$$

and

$$CO_2 + 4H_2 \rightarrow CH_4 + 2H_2O \qquad (9)$$

are both exothermic, *i.e.* require cooling to proceed towards the right and are assisted by increased pressure. Both external and internal water cooling have been applied, although the presence of steam tends to reverse the reactions.

In order to start the reaction at a low temperature highly active catalysts are used which tend to be deactivated by contaminants, particularly sulphur compounds present in the gas. If derived from impure feedstocks the reactant gases have therefore to be purified before they contact the methanation catalyst.

In view of the high exothermicity of the reaction and the temperature-dependence of the equilibrium, shown in Table 5.2, it is as a rule easier to carry out the reaction in several steps, restricting conversion and therefore heat evolution in each instance. The products are then cooled before further methanation is allowed to take place. Dilution of the reactants with an inert non-reacting gas to reduce conversion rate and improve cooling is an alternative.

5.8 CONCLUSIONS

The gasification of liquid and solid fuels on a commercial scale to produce SNG is based on a complex series of chemical reactions. While in different routes to SNG certain reactions predominate and therefore some chemical equilibria are of greater importance than others—e.g. the reactions of carbon with steam in reforming processes and with hydrogen in hydrogasification—it seems likely that in practice in all gasification processes the respective equilibria mentioned in the previous fourteen reactions will become established.

It follows that there is considerable similarity between the different commercial routes as far as chemical equilibria are concerned, and differences, if any, are due to incomplete equilibrium and speed of reaction which are a function of catalysts used, and since the latter tend to be sensitive to contamination, of impurities present in the feedstock.

The large number of process routes leading from different natural fossil fuels to SNG, which will be discussed in the next few chapters, are an indication of the great ingenuity that has been employed in optimising reaction rates and equilibria to obtain maximum yields of SNG with a minimum expenditure of energy, while using different hydrocarbon raw materials of varying complexity and degree of contamination.

REFERENCES

1. American Gas Association (1969). *Gas Engineer's Handbook*, Industrial Press Inc., New York.
2. Waterman, W. W. (1973). *Introduction to SNG Processes*, Paper 7, Inst. Gas Tech., SNG Symposium I.

3. Goldstein, R. F. (1958). *The Petroleum Chemicals Industry*, 2nd Ed., E. & F. N. Spon, London.
4. B.P. Trading Ltd (1972). *Gas Making and Natural Gas*, B.P. Trading Ltd, London.
5. Morgan, J. J. (1953). *Gasification of Hydrocarbons*, Moore Publishing Co., New York.
6. Meunier, J. (1958). *Gazéification et oxydation des combustibles*, Masson, Paris.

Chapter 6

Naphtha Gasification

6.1 INTRODUCTION

While practically all gasification processes can make use of naphtha as a feedstock there are certain processes for which it is essential to operate on a feedstock light enough to be completely desulphurised, preferably by one of the established purification routes. Such hydrocarbon feedstocks comprise LPG, natural gas condensate, light and wide-range naphthas and, for certain types of gasification processes, also heavy naphtha.[4]

Gasification routes that typically require a naphtha or lighter feedstock include the various steam reforming processes and one version of the British Gas Corporation gas recycle hydrogasification (GRH) process, although the latter lends itself to conversions of materials heavier than naphtha as well and therefore will be discussed under the general heading of gas oil and middle distillate gasification (Chapter 7).

The conversion of naphtha into a gas by means of steam can take place essentially by two distinct routes, depending on reaction temperature, which lead either to a mixture of carbon monoxide and hydrogen at high temperatures ($c.$ 800 °C), or to methane diluted with a certain amount of carbon dioxide and hydrogen at about 470 °C. The latter route which is known as low temperature reforming, is preferable if SNG is to be produced, since under those circumstances the subsequent steps of gas purification and enrichment can be simplified. In fact, if the SNG only has to meet technical interchangeability criteria it may be sufficient to remove most of the carbon dioxide, when the resultant gas acquires similar combustion characteristics—though not the same calorific value—as those of natural gas. Therefore, if commercial interchangeability is not essential, it can be left in this form. Only if both technical and commercial compatibility are specified does further treatment become essential.

The use of a high temperature reforming or other process, on the other hand, necessitates fairly complex multi-step methanation or a similar type of enrichment step. The gas as produced consists mainly of low calorific value components (carbon monoxide and hydrogen) and is clearly unsuitable for natural gas replacement.

The advantage of reforming processes over hydrogasification has already been touched upon. It is essentially the very large hydrogen requirement of the latter and the subsequent need to build the disproportionately large hydrogen plant needed for hydrogasification, that favours the economics of a reformer. The latter needs only hydrogen in relatively small quantities for desulphurisation of the feedstock and possibly for the saturation of any olefins present. The majority of SNG plants in operation are therefore of this type.

There are a number of fully proven low temperature reforming technologies, and at least three types of commercial naphtha to SNG reforming plants have been built over the last few years. These are the British Gas Corporation's 'Catalytic Rich Gas' (CRG) plants, the German Lurgi Company's 'Gasynthan' plants and the Japan Gasoline Company's 'Methane-Rich Gas' (MRG) process. It is proposed to discuss each of these in some detail in this chapter and to provide economic information on them in Chapter 11.

6.2   THE BRITISH GAS CORPORATION CATALYTIC RICH GAS (CRG) PROCESS[1,8,14]

The first process to produce a natural gas substitute from naphtha and similar feedstocks originated at the Solihull laboratories of the British Gas Corporation. It was based on development work done in the 1960s aimed originally at the manufacture of a typical European town gas, i.e. of a mixture of methane, hydrogen and inerts, of a calorific value of around 4500 kcal/m^3 and a Weaver flame speed of about 42 (Delbourg combustion potential 30).

When it became necessary to produce a richer gas additional stages of enrichment, on the one hand by removing carbon dioxide, on the other hand by methanating the remaining carbon oxides present, were added to the original CRG town gas plants of which, by the time the town gas era ended in Britain and elsewhere—mainly owing to the introduction of natural gas—there existed some 40 units in the UK, Japan and several European countries.

The principle of the CRG catalytic low temperature reforming process lies in a highly active nickel catalyst that converts a sulphur-free hydrocarbon feedstock and steam, supplied in the ratio of about 1 to 2, into a mixture of methane, carbon dioxide, and a small amount of hydrogen.

The desulphurisation system used in the CRG process is based on the conversion of organic sulphur compounds into H_2S, which is subsequently removed from the gas stream by means of a chemical absorbent.

For this conversion of organic sulphur into hydrogen sulphide, a reactor is filled with either Nimox (nickelmolybdate) or Comox (Cobaltmolybdate) catalyst. The feedstock is vaporised, mixed with a recycle gas containing mostly hydrogen, and preheated to about 350 °C. The mixture is passed over the hydrofining catalyst and is split into naphtha vapour and hydrogen sulphide, the H_2S being removed by adsorption on, and reaction with, a bed of zinc oxide. The latter is thereby gradually converted into zinc sulphide and clearly its capacity for sulphur absorption must therefore be limited.

It is common practice to operate H_2S absorption as a batch process, i.e. to change beds, or open and refill the reactor with fresh zinc oxide when the old charge is saturated. However, it will be appreciated that this is only feasible if the sulphur content of the feedstock is low. For higher sulphur contents it is preferable to remove the bulk of the sulphur in a separate hydrofining plant—these operate with liquid regenerable absorbents for H_2S and therefore have no particular limits on sulphur content—and to use the partially desulphurised material as feedstock. Alternatively, sulphur can be reduced in one operation to the level required for the CRG reforming step, about 0·2 ppm, in a separate plant and the feedstock preparation section can be eliminated from the CRG plant. However, the clean feedstock must be guarded carefully against any form of contamination.

At the exit from the H_2S absorber steam is added to the desulphurised feedstock in proportions of between 1·6 and 2·2 by weight, the exact ratio being a function of preheat temperature, reforming pressure and condition of the catalyst. The mixture is now raised to reaction temperature (between 400 and 510 °C) and introduced into the reforming reactor, where a complex series of intermediate reactions converts the feed, steam and residual recycle hydrogen, in accordance with the overall reaction—written here for normal hexane rather than a full-range naphtha:

$$4C_6H_{14} + 10H_2O + xH_2 \rightarrow 19CH_4 + 5CO_2 + yH_2$$

The complex character of the reactions that take place in a CRG converter is underscored by the unusual temperature distribution in the catalyst bed. When the catalyst is fresh the inlet temperature, say 450 °C, remains unchanged over the first 25–30 cm of the bed; at that level it rises fairly suddenly by some 50 °C indicating the onset of an exothermic reaction. After a few weeks operation of the catalyst it may become necessary in the interest of complete conversion to increase the inlet temperature by a few degrees. At that stage the constant temperature range near the inlet will be about the same but a slight temperature dip will have developed where originally the temperature rose, and the rise to the original temperature will now have receded downwards by a few centimetres. As the catalyst becomes less active the inlet temperature has to be raised further and the temperature dip becomes more pronounced. It also moves down the catalyst bed, whose degree of deactivation can be followed quite clearly with the aid of a temperature trace.

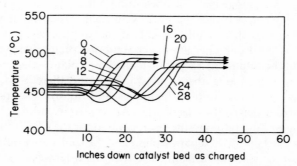

FIG. 6.1   Temperature distribution in the CRG reactor at Portslade Works, Brighton. Numbers refer to running time in weeks. (Courtesy: British Gas Corporation.)

The series of temperature profiles shown in Fig. 6.1 is indicative of an endothermic reaction followed rapidly by an exothermic one and could be due, for example, to the formation of carbon monoxide and hydrogen in the first instance and their subsequent recombination to form methane. However, other interpretations are also possible. The addition of hydrogen, for example, results in a reduction or even elimination of the temperature dip, and the final temperature further down the bed is raised by about 15 °C. The recycling of product gas mixture, *i.e.* methane/carbon dioxide/hydrogen, and a simultaneous reduction in steam permits a lowering of the inlet temperature from 450 °C to almost 350 °C and a complete change in the character of the overall reaction (a phenomenon to be discussed under the heading of hydrogasification). It also eliminates the temperature dip on entry, probably owing to an exothermic feedstock hydrogenation step balancing the endothermic reforming processes. The overall temperature increase in the catalytic hydrogasification variant of the CRG reaction is, furthermore, raised from about 50° to over 100 °C.[7]

The gas produced in a CRG reactor with an inlet temperature of 450 °C or higher is not interchangeable with methane or other natural gases consisting mainly of methane. Table 6.1 shows its composition and combustion characteristics as it emerges from the reformer and after a series of subsequent enrichment treatments carried out in one of the following series of steps:

(a) Methanation I + Methanation II + CO_2 removal.
(b) Hydrogasification + Methanation + CO_2 removal.
(c) Propane blending.
(d) CO_2 removal and propane blending.

The series of steps indicated in (a) and illustrated diagrammatically in Fig. 6.2 produces a gas of satisfactory quality, being interchangeable with natural gases of high methane content (*see* Chapter 3). The gas has to be cooled to about 280 °C both before the first methanator and after its temperature rise due to methanation to

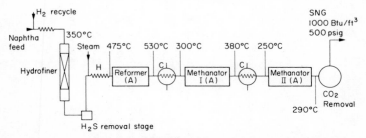

FIG. 6.2.   The CRG double methanation process. A, nickel catalyst; C, intercooler; H, heater.

TABLE 6.1
British Gas Corporation CRG Process
Gas Composition and Combustion Characteristics

Gas composition (% vol)	Outlet CRG reactor (dry)[a]	Methanation I	Methanation II	ex CO$_2$ absorber (methanation)	ex CO$_2$ absorber (hydrogasification)	Propane blending No CO$_2$ removal	Propane blending CO$_2$ removal
CH$_4$	65·5	71·6	76·7	98·5	98·3	54·2	73·2
H$_2$	11·8	6·2	0·6	0·9	1·05	9·7	18·0
CO	0·8	0·1	0·1	0·1	0·15	0·1	0·1
CO$_2$	21·9	22·1	22·6	0·5	0·50	18·0	0·5
Propane	—	—	—	—	—	18·0	8·2
Heating value, gross, Btu/scf, 15°C, dry	691	739	781	1 000	1 000	1 000	1 000
Relative density, 15°C (air = 1)	0·706	0·734	0·771	0·557	0·555	0·809	0·548
Wobbe Number, Btu/scf, 15°C, dry	823	862	889	1 342	1 342	1 100	1 325
Flame speed factor (Weaver)	19·6	15·3	12·2	14·2	14·4	17·1	19·7

[a] Feedstock C/H 5·9 wt/wt; overall steam/naphtha 1:2; pressure 35 atm.

ensure a satisfactory gas quality at the exit; water content is generally reduced to about 0·1% wt before the gas enters the second methanator.[5] (Catalysts and operating conditions of methanators are discussed in greater detail in Chapter 10.)

Instead of methanation in two stages the CRG exit gas can also be enriched and made interchangeable with methane gas by catalytic hydrogasification or by a combination of hydrogasification and methanation[14] (see Chapter 7 and Fig. 6.3).

Table 6.1 also indicates the properties of a rich gas obtained by blending CRG reactor exit gas with propane (c). This alternative increases the heating value to the required level but does not produce a fully interchangeable gas, since without prior CO_2 removal (d) the specific gravity is too high and therefore the Wobbe Index too low. After CO_2 removal however the flame speed factor then becomes higher than desirable. While propane enrichment as such therefore is not an acceptable route to SNG it can, if necessary, be combined with other forms of after-treatment. Partial enrichment by methanation or hydrogasification followed by propane enrichment or trimming provides an opportunity for accurate adjustment of properties and is at times the most economic route to SNG.

An aspect of the CRG process that has not been discussed so far is the need for, and production of, recycle hydrogen for the hydrofining and desulphurisation of the feedstock. This pretreatment is carried out in a subsidiary plant in which a side-stream of the product gas is taken from the CRG reactor outlet in the case of double methanation, or from the hydrogasifier outlet in the case of hydrogasification/methanation. The gas is subjected to high temperature reforming by first mixing with steam in a weight ratio of about 6:1, then passing over a catalyst in a fired tubular reactor, where the temperature reaches 800 to 850 °C. The effect of the catalyst, usually of low nickel content (5–8%), is to convert the methane into carbon monoxide and hydrogen. The gases are then cooled and passed, without prior steam removal, over a 'shift' catalyst, which can be either an iron-chrome high temperature catalyst operating at 350–500 °C, or one of the newer catalysts such as the Girdler/Du Pont material which is effective at 200–300 °C.

The effect of the 'shift' process is to convert carbon monoxide into hydrogen, thus increasing the concentration of the latter from 35 to 45% vol at the reformer outlet to 70–75% vol after the shift converter. Carbon dioxide present in the feed gas and additional CO_2 formed

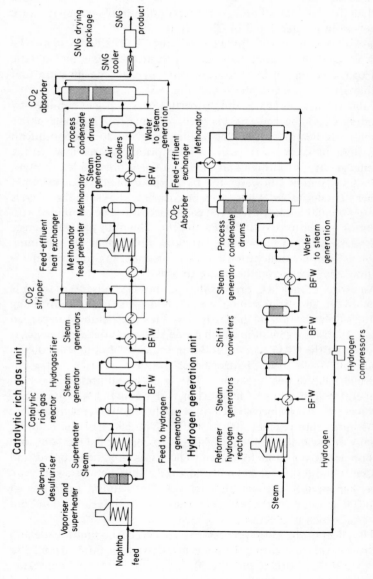

FIG. 6.3 Schematic flow diagram, synthetic natural gas plant, CRG-hydrogasification process. (Courtesy: British Gas Corporation.)

during the shift reaction is then removed by amine or potassium carbonate scrubbing, and an almost pure hydrogen stream is recycled after final methanation (to remove traces of carbon oxides) and mixed with the naphtha feed at the feed preheater inlet.

6.3 THE LURGI GASYNTHAN PROCESS[9]

Not unlike the British Gas Corporation's CRG process, the Gasynthan route was developed from a gasification process originally conceived to produce a gas of about 4500 kcal/m^3 as a replacement for coal gas and similar town gases that were distributed at the time in Germany, the rest of Europe, Japan and a number of other areas. Under the name of Recatro, Lurgi, in co-operation with BASF, developed a catalyst and process that converted light hydrocarbons up to and including wide-range naphtha into a gas of a composition very similar to that of the first stage of the CRG process. Again, very much like their British counterparts, Lurgi also developed methanation, hydrogasification and carbon dioxide removal stages that converted the intermediate calorific value gas into a rich gas of properties closely resembling those of methane-rich natural gases.

The principal difference between the various commercial gasification processes lies in the proprietary catalysts used in the first gasification stage. In the case of the Gasynthan process, this is again a nickel catalyst of very intense activity and consequently highly sensitive to contamination by sulphur compounds, halogens, oxygen, lead (derived from tetraethyl lead in gasoline) and others.

Owing to this high activity, the inlet temperature for the steam/naphtha vapour blend can be reduced to 400 °C, and it is also claimed that the minimum permissible steam/naphtha ratio used by Gasynthan is lower than that of competitive processes, since a lower temperature implies a more favourable equilibrium in regard to carbon formation. However, most Gasynthan plants are believed to work in the close neighbourhood of 2:1 steam to carbon ratio, i.e. the same as for competitive processes, possibly because the risk of permanent deactivation of the catalyst by carbon deposition is even greater owing to its high reactivity, and also bearing in mind that excess steam absorbs reaction heat and prevents a possible temperature run away with consequent carbon deposition and blocking of the reactor.

The feedstock vaporisation system of the Gasynthan process differs from that of other commercial low temperature reformers in using heat exchange between hot product gas and liquid feed, rather than direct firing, to vaporise the feedstock. By not exposing liquid feed at high pressure to direct flame impingement and the possibility of local over-heating, it is claimed that the likelihood of cracking, carbon deposition and ultimate blockage of feed preheaters is much reduced, if not eliminated. By this means it is possible to extend the boiling range of the feedstock into the heavy naphtha and kerosine range without any danger of thermal cracking.

The desulphurisation stage of the Gasynthan process is identical in conception with that of the CRG route. Recycled hydrogen and naphtha vapour are passed through a reactor packed with cobalt or nickel molybdate catalyst. Hydrogen sulphide is formed from sulphur-containing organic molecules and is collected by deposition on a bed of zinc oxide or Luxmasse contained in a second reactor. The entire desulphurisation system is maintained at 330–360 °C, both hydrodesulphurisation and sulphur absorption being slightly exothermic and the overall effect being one of an adiabatic system. Again for high sulphur content feedstocks external hydrofining facilities, either complete or partial, can be provided.

Any olefins present are, of course, also hydrogenated under these conditions and since this is an even more exothermic reaction their concentration in the feedstock must be restricted, generally to less than 5 %. The degree of hydrogenation of aromatics, on the other hand, is limited and usually takes place in the low temperature reforming stage. It has been suggested that naphthenes and aromatics in the feed should not exceed 50 points, where one point is awarded for each volume percentage of naphthene and two for aromatics.

The Gasynthan low temperature reformer is characterised by a similar type of temperature profile with trough and sudden rise, both originally at the point of entry and gradually travelling down the reactor as the catalyst becomes deactivated just as applies in the CRG system. However Gasynthan reactor temperature profiles at increasing catalyst age show a widening—rather than displaced—endothermic zone and a gradual reduction in the temperature rise during the exothermic phase, indicating a somewhat different reaction pattern from the one in the CRG reactor (see Fig. 6.1).

In both systems components of the gas leaving the low temperature reformer stage appear to have reached chemical equilibrium and further formation of methane can only be achieved by introducing a further component or lowering the temperature. Gasynthan at present uses catalytic hydrogasification for enrichment, *i.e.* both reduces the temperature to about 350 °C and introduces additional purified naphtha vapour to react with the residual hydrogen and steam. The temperature profile in the second reactor, however, rises from the very beginning since at the lower temperature no endothermic cracking or reforming occurs, and the surplus hydrogen ensures the immediate start of exothermic hydrogenation reactions.

Again, as in the CRG route, it is desirable to correct the combustion characteristics of the gas at that point by a methanation stage.* This ensures the removal of any residual hydrogen, and after subsequent absorption of the bulk of the carbon dioxide present in the gas, the final product becomes completely interchangeable with methane-rich natural gas. Discharge pressure is usually about 35 atmospheres gauge.

Table 6.2 lists the composition and combustion properties of Gasynthan SNG as it leaves the methanator, and following the final CO_2 removal step for the two cases of methanation with and without prior water removal.

* This sequence of processing steps is representative of the so-called 'advanced Gasynthan' process.[10] Other possibilities have been described, *viz.* single stage gasification followed by one or more methanation steps (standard Gasynthan),[11] standard Gasynthan with more highly active catalyst, and single stage gasification with one methanation step and recycle of low CO_2 product gas.[10]

TABLE 6.2
Gasynthan Product Gas—Composition and Properties

Composition (% vol)	Wet methanation (% vol, dry)		Dry methanation (% vol, dry)	
	ex Methanator	*ex* CO_2 removal	*ex* Methanator	*ex* CO_2 removal
CH_4	76·07	97·32	76·81	98·08
H_2	1·29	1·66	0·56	0·72
CO	0·02	0·02	0·08	0·10
CO_2	22·62	1·00	22·55	1·10
Heating value, kcal/m³	775·6	992·3	780·9	998·2
Spec. gravity (air = 1)	0·768	0·556	0·771	0·560
Wobbe number	884·9	1 331·0	888·9	1 331·1
Weaver flame speed factor	12·4	14·4	12·2	14·2

Hydrogen for feedstock preparation and hydrodesulphurisation is prepared in a separate plant section that processes product gas from the hydrogasification stage in a high temperature steam reformer. Passing gas and steam over a catalyst in an externally fired tube converts the mixture into carbon monoxide and hydrogen. Further addition of steam and cooling completes the shift to hydrogen and

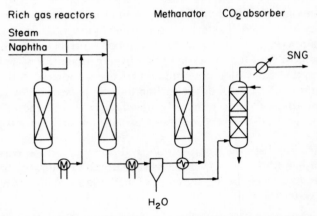

FIG. 6.4 Simplified flow diagram for Advanced Gasynthan process (two-stage gasification with one-stage methanation). (Reprinted from *Hydrocarbon Processing*, **52**, No. 1, 93, Jan. 1973. Copyright Gulf Publishing Co., all rights reserved.)

carbon dioxide, and absorption of the latter in an amine solvent results in almost pure hydrogen which is recycled to the inlet of the naphtha desulphurisation system.

A simplified flow diagram for the Advanced Gasynthan SNG system is shown in Fig. 6.4.

6.4 THE JAPAN GASOLINE CO., METHANE-RICH GAS (MRG) PROCESS[2,3,6,12,13]

The Japan Gasoline Co.'s MRG process was developed by JGC in co-operation with its chemical affiliate Nikki Chemical Company, the developers and suppliers of the appropriate catalyst. Again a number of low calorific value gas plants had been designed and built in Japan to produce town gas, synthesis gas and enrichment gas from naphtha and LPG before SNG production was considered.

While the MRG town gas plants in Japan were relatively small, the biggest produced 14 million scfd per stream, the SNG facilities that are now being built by Japan Gasoline Co. and its licensee in the United States are very much larger (150–250 million scfd). A complete redesign of the process equipment, additional pilot plant and engineering studies by JGC and their US licensee UOP were consequently undertaken. The outcome was a gasification system somewhat different from the previously discussed CRG and Gasynthan processes.

The MRG route from naphtha to SNG consists of the usual four stages: hydrodesulphurisation, gasification, enrichment and carbon dioxide removal. The reforming catalyst differs from those used in the CRG and Gasynthan processes in its ability to deal with some sulphur contamination. It appears that minor breakthroughs of sulphur compounds would deactivate only the surface layer of the catalyst bed. A catalyst designated N-185, which in addition to nickel contains copper and chromium, is claimed to retain sulphur extremely firmly so that deactivation does not spread throughout the bed. It also maintains sufficient activity to prevent the feedstock cracking to carbon before gasification reactions are initiated.

One can consequently use desulphurisation systems other than the Comox or Nimox combined with zinc oxide or Luxmasse employed in the previously discussed routes. The Hydrobon process, developed by UOP for instance, is a naphtha to gasoil hydrodesulphurisation system in which naphtha vapours and extraneous hydrogen are passed over a desulphurisation catalyst that decomposes organic sulphur compounds. The vapour is cooled, condensed and separated from the recycle gas; residual hydrogen sulphide in the condensed naphtha is removed by fractionation.

The desulphurised product, particularly if it is high boiling (say 260 °C end point), will contain up to 5 ppm of sulphur and would thus be unsuitable as a feedstock for other low temperature reforming processes. However, in the MRG process, catalyst sensitivity to sulphur compounds is reduced both by the nature of the catalyst and by the fact that extraneous hydrogen is once more injected ahead of the feedstock preheater together with the steam and naphtha.

The mixture of naphtha vapour, steam and hydrogen enters the upper catalyst bed of the adiabatic low temperature reforming reactor at a temperature of about 450 °C. In order to ensure optimum utilisation of the steam the effluent from the first bed is blended with additional naphtha vapour and re-equilibrated over a second bed, essentially producing a hydrogasification sequence without, however, the temperature drop between reactors common in CRG and Gasynthan plants. There follows a further lower temperature methanation, again over the original MRG catalyst, which reduces the hydrogen content of the gas from almost 10 % vol to about 3 % vol. Further hydrogen reduction requires 'dry' methanation, *i.e.* water removal from the gas by condensation, before it enters the second methanator.

The final processing step, carbon dioxide removal, is preceded by further cooling and water separation, before the gas enters the carbon dioxide absorber. The latter is basically of the same type used in other SNG processes, *i.e.* uses an amine solution to absorb the sour gases. The saturated solvent is subsequently reheated and the absorbed CO_2 is stripped in a separate column so as to permit the recovered lean solvent to be re-used. The carbon dioxide content of the gas is reduced from about 17 % vol to between 0·5 and 1 % vol.

Table 6.3 lists operating conditions, feedstock characteristics, gas quality and yield of a typical MRG plant for the conversion of a high sulphur content wide-boiling naphtha into SNG.

Figure 6.5 is a simplified flow-sheet of the MRG process, in which it is assumed that extraneous hydrogen is supplied to both hydrodesulphurisation and low temperature reforming stages. While this is, in fact, a possible method of operating it will not apply in a situation where a large SNG plant stands in isolation and hydrogen would have to be supplied in tanks or cylinders. Under those conditions part of the hydrogasifier exit gas would have to be steam reformed at a higher temperature, as in the CRG and Gasynthan processes, and submitted to shift conversion to increase its hydrogen content and its carbon dioxide component would have to be reduced or eliminated by amine absorption.

In addition to the three commercial low temperature steam reforming routes there are a number of process developments which, so far, do not appear to have reached the commercial plant stage. Among these is the Adtec process, developed at the Chicago Institute of Gas Technology. IGT appear to have combined known desulphurisation, steam reforming and methanation technology, originating both within their laboratories and outside, into an undoubtedly workable naphtha gasification process. Phillips Petroleum Co. have also developed a process named Synnat,

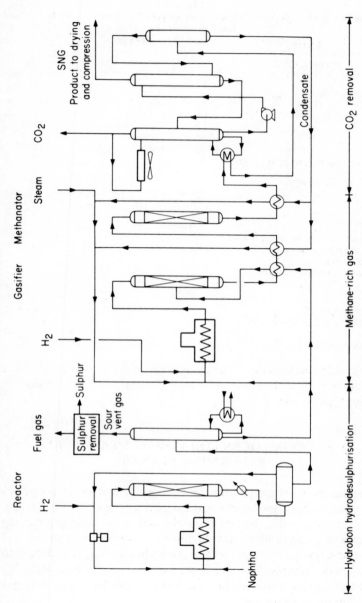

Fig. 6.5 MRG Process for converting naphtha to SNG. (Source: Erickson, R.A. Paper to IGT's SNG Symposium (March 1973).)

TABLE 6.3

Japan Gasoline Co. Methane-Rich Gas Process Operating Conditions, Yields, Feed and Gas Characteristics

Plant capacity, million scfd		125	
Delivery pressure, atm.		34	
Naphtha charge rate, bsd		23 900	
Naphtha properties			
Boiling range, IBP	32·2	Hydrocarbon types,	% vol
°C 10%	60·0	Paraffins	74
50%	104·4	Aromatics	7
90%	157·2	Olefins	1
EP	185·0	Naphthenes	18
Sulphur content, wt ppm 1 000			
SNG product quality			
Analysis, % vol CH_4	98·00		
H_2	1·50		
CO	0·01		
CO_2	0·49		
Heating value, gross (dry), Btu/scf		995	
Specific gravity, air = 1		0·553	
Wobbe Number		1 338	
Weaver flame speed factor		13·9	
By-products			
Sulphur, tons/sd	2·7		
Carbon dioxide, million scfd	33		

From Refs. 2, 6, 12, 13.

applicable particularly to the conversion of LPG into a substitute natural gas. No orders have so far been placed for full scale versions of either type of plant.

6.5   FEEDSTOCKS FOR LOW TEMPERATURE REFORMING PLANTS

The concept of low temperature reforming is almost invariably tied to the gasification of feedstocks in the naphtha or light distillate range. While it is possible, as has been claimed by various engineering contractors and catalyst manufacturers, to convert light gas oil and sometimes even heavier gas oil and crudes by catalytic reforming into either low calorific value or even high calorific value gases, all commercial processes so far use either naphtha or natural gas condensate as feedstock.

On the other hand, there is no technical reason why feedstocks lighter than naphtha should not be fed to low temperature reformers. Wherever this is economical and feasible (and it will be appreciated that normally this is not the case), propane, butane, mixed LPG, and even ethane can be processed in low temperature reforming plants without substantial changes in design. As a rule, however, supplemental natural gas from these lighter feedstocks would be prepared by simpler means, *i.e.* air dilution or dilution with lean combustible gases.

With regard to naphtha feedstocks, it is worth remembering that they can differ both in average and maximum boiling points and also in regard to sulphur and impurities content. Process limitations usually apply to maximum boiling point, although this is not strictly logical within the limits of attainable boiling point anyway, since process design and operating cost are usually a function of average feed properties rather than permissible extremes. In particular, heavy feedstocks will produce more carbon dioxide and therefore will be more expensive to process. Simultaneously they will tend to deactivate the reforming catalyst more rapidly and therefore require more frequent catalyst changes. Neither of these disadvantages will become apparent if the feedstock contains very small concentrations of high boiling material, *i.e.* if its average boiling point is low despite its end boiling point being high (as is often the case with natural gas condensates, *see* Chapter 4).

Different feedstock end points have nevertheless been specified by the various designers of commercial plant. For CRG low temperature reformers, for instance, the maximum permissible feedstock boiling point for plants with methanation facilities is 185 °C. Hydro-gasification feedstocks, on the other hand, have until recently been restricted to a maximum of 150 °C. Fewer limitations have been imposed in the case of the Gasynthan process, and a 200 °C end point naphtha, although allegedly more expensive in processing cost and somewhat inferior in gas yield, can be processed in standard gasification plant of this type. It has similarly been claimed by Japan Gasoline Company that their catalyst and low temperature reformers are not sensitive to feedstock end point and that most heavy and wide range naphthas can be processed in these plants.

The limits of feedstock sulphur content are less well defined. It has been mentioned that thorough feedstock desulphurisation is essential, whatever the type of process. Process limitations therefore refer to economics rather than technology, and higher sulphur content must simply be allowed for in the design of the plant. Specifically, if the average sulphur content is above about 200 ppm, it will become advantageous to remove the bulk of the sulphur present in the feedstock in a separate plant in order to avoid overloading the zinc oxide or Luxmasse sulphur absorption system. Once it has been decided to desulphurise the feedstock in separate facilities, there are no longer any limits on sulphur content and any restriction based on feedstock can be considered in an economic context; clearly removing 1 % wt of sulphur will be more expensive than removing, say, 200 ppm.

The aromatics content of low temperature reforming feedstocks is less critical than it would appear. Most reformers can process feedstocks containing between 7 and 10 % vol, of aromatics, without any reduction of catalyst life or performance. However, it will be appreciated, with reference to Chapter 4, that such high aromatics contents occur very rarely in naphtha obtained from the vast majority of crude oils. The effect of naphthenes is of course similar to that of aromatics, although any deactivation effects that they might produce are balanced by the simultaneous formation of hydrogen when naphthenes are cracked to aromatics and gas in the initial stages of a low temperature reforming process.

The presence of lead, which could be derived from the tetraethyl lead anti-knock agent present in naphtha based gasoline is highly undesirable since, it is claimed, it will deactivate the nickel reforming catalyst. However, in plants with preliminary hydrodesulphurisation of the feedstock, all lead compounds will have been absorbed on the first stage catalyst beds, and it would seem unlikely that they could reach the reforming catalyst unless present in such massive concentrations that the first catalyst bed became saturated. The dangers of lead poisoning may therefore have been overrated. Other impurities such as chlorine and other halogens appear to have little or no effect on hydrofining or reforming catalyst activity if present in small concentrations (less than 20 ppm). However, contamination by such impurities, including lead, to any significant degree, is unlikely.

6.6 CONCLUSIONS

Low temperature catalytic reforming is the basis of a series of commercial SNG processes which use naphtha or similar light hydrocarbon feedstocks.* A number of large commercial SNG plants based on at least three designs are now operating in the US and further plants are under construction. The processes used thus appear to be technically efficient and economically attractive. Typical design plant capacity is about 250–500 million ft^3 of SNG, and with 100 million cubic feet of SNG requiring about 2500 tons of feedstock, the daily naphtha demand would be of the order of 12 500 tons for the larger plant.

Unfortunately, however, the supply of naphtha type feedstocks to match this very large demand is strictly limited. Not only have naphtha prices risen in recent years even more than those of other petroleum products, but contractual supplies for new plants have on occasions simply not been negotiable. Interest in future construction of naphtha-based gasifiers has therefore much declined.

* For convenient reference, the composition and properties of the finished SNGs are compared with those of the product gas from other SNG manufacturing processes in Appendix A.

REFERENCES

1. Davies, H. S., Lacey, J. A. and Thompson, B. H. (1971). *Processes for the Manufacture of Natural Gas Substitutes*, Gas Council Res. Comm. GC 155.
2. Erickson, R. A. (1974). *JGC Methane Rich Gas Process*, Paper 9, Inst. Gas Tech., SNG Symposium I.
3. Fukunaga, K., Nojima, S., Okagami, A. and Ward, D. J. (1973). *New Methanation Catalyst*, 5th Synth. Pipeline Gas Symposium, Chicago.
4. Lom, W. L. and Agius, P. J. (1975). *Technology and Economics of Clean Fuel Gas Manufacture from Liquid Petroleum*, 9th World Petroleum Congress, Tokyo, P.D. 17, Paper 1.
5. Long, G. (1972). Why methanate SNG? *Hydrocarb. Process.*, **51**, 8.
6. Morikawa, K., Nojima, S. and Okagami, A. (1975). *Technology and Economics of SNG Manufacture from Naphtha*, 9th World Petroleum Congress, Tokyo, P.D. 17, Paper 2.
7. Tittle, R. W. and Hartley, W. (1974). *The Potential of the GRH in the Production of SNG*, Paper 12, Inst. Gas Tech., SNG Symposium I.
8. Weiss, A. J. (1974). *The Catalytic Rich Gas Process*, Paper 8, Inst. Gas Tech., SNG Symposium I.
9. White, G. A. (1974). *The Gasynthan Process, a BASF/Lurgi Development*, Paper 10, Inst. Gas Tech., SNG Symposium I.
10. Jockel, H. and Triebskorn, B. E. (1973). Gasynthan process for SNG, *Hydrocarb. Process.*, Jan., 93.

11. Baron, G. and Hiller, H. (1967). *Erdöl u. Kohle*, 20(3), 196.
12. Thornton, D. P. *et al.* (1972). MRG process for SNG, *Hydrocarb. Process.*, Aug., 81.
13. Ward, D. J. (1973). SNG by the MRG process, *Gas*, **49**(5), 67.
14. Hebden, D. and Stanton, C. H. (1973). *Recent Developments in the Production of a Substitute Natural Gas by the CRG Process*, Paper to 12th World Gas Conference, Nice, IGU/B-4.

Chapter 7

Hydrogasification of Distillates and Heavy Liquid Fuels

7.1 INTRODUCTION

The gasifying medium considered so far for the commercial production of SNG was essentially steam. The thermally almost neutral low temperature steam reforming reaction permits the direct conversion of a hydrocarbon of up to about C_8 carbon number into SNG. Heavier hydrocarbons than naphtha have not so far been reformed in commercial equipment, there being problems of complete desulphurisation of higher boiling hydrocarbons and the consequent reformer catalyst poisoning on the one hand, and the tendency of these hydrocarbons to decompose by pyrolysis and thereby to deposit elemental carbon, on the other.

It has already been mentioned that both catalyst deactivation and carbon-forming tendency can be mitigated by the presence of hydrogen in the feed. It has also been noted that reformer inlet temperature can be reduced if hydrogen enters simultaneously with the desulphurised feedstock and steam. It follows therefore that a combination of hydrogasification and steam reforming could be an interesting alternative to the direct low temperature reforming of naphtha hydrocarbons. Direct hydrogenation of thermally unstable hydrocarbons, *i.e.* contact between feedstock and hydrogen with no intervening reforming stage, is also a possibility provided means of dealing with deposited carbon are available.

Before discussing the hydrogasification of naphthas with low temperature steam reformer gas as well as the direct hydrogenation of higher hydrocarbons in any great detail, some consideration is given to the equipment available for hydrogenating vaporised and liquid hydrocarbons. In the following section of this chapter it is therefore proposed to describe hydrogenators for light and heavy feedstocks. In subsequent sections we shall discuss hydrogasification of light

115

feedstocks such as naphtha and kerosine and the hydrogenation of middle distillates, crude oils and residual fuels; and in a final section some thought will be given to the manufacture of hydrogen required for various processes of SNG manufacture.

7.2   HYDROGENATOR DESIGN[2,6]

The reaction between hydrogen and a hydrocarbon is exothermic, as discussed in Chapter 5, however, its activation energy, which varies between different hydrocarbons, is always fairly high. The initiation temperature is in fact of the order of 700 °C and it is therefore necessary to preheat the reactants to that temperature. Since most hydrocarbons, particularly higher molecular weight hydrocarbons, undergo a fair amount of cracking if heated to such a high temperature, ways and means have to be found to

—dilute the feed with hydrogen during preheating, or
—heat very rapidly to minimise pyrolysis, or
—provide deposition surfaces for equilibrium carbon.

While the first two methods rely on the slowing down of the rate of carbon formation, the last is based on acceptance of equilibrium carbon, which at the mean hydrogenation temperature will be both formed and removed in accordance with a reaction of the type:

$$C + 2H_2 \rightleftharpoons CH_4$$

Dilution with hydrogen and rapid heating on the other hand is achieved in the Gas Recycle Hydrogenator (GRH), designed by the Solihull Research Station of the British Gas Corporation, and shown diagrammatically in Fig. 7.1. The GRH reactor consists of a high pressure cylindrical steel shell with an injection nozzle for the reactants mixture at the top, a cylindrical baffle for product gas recycle, a product outlet near the top and a drain for condensed liquid products near the bottom. Hydrogen gas and hydrocarbon vapour are injected into the central space, descend to the bottom of the reactor and return through the outside annulus. Part of the material is recycled internally, *i.e.* does not leave the reactor after the first pass; however, none leaves before completing at least one passage through core and ring.

Conversion in the GRH reactor takes place under adiabatic

conditions and the vessel is carefully insulated on the outside with the result that product that enters at about 450 °C is rapidly raised to about 750 °C, the operating temperature of the process and the temperature of the gas as it leaves the reactor. Residence time of hydrocarbon material in the reactor is of the order of 10 sec.

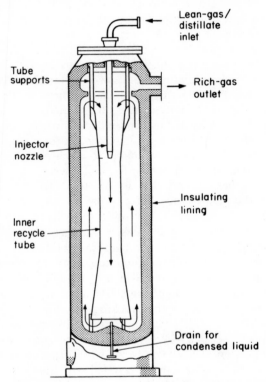

FIG. 7.1   A commercial version of the gas-recycle hydrogenator. (Courtesy: British Gas Corporation.)

Operating pressures of as low as 10 atm and as high as 100 atm have been used successfully. However, rate of hydrogenation being a function of roughly the second power of hydrogen partial pressure, there is considerable incentive to operate the plant close to its higher rather than lower pressure limit. Also, since minimum residence time is absolute and not a function of pressure, the capacity of high pressure gasifiers is greater than that of low pressure units.

A further important consideration is hydrogen purity. Clearly, since it is the partial pressure of hydrogen rather than the absolute pressure that accelerates the reaction, the highest possible hydrogen concentration is required. Hydrogen purification by means of high temperature steam reforming of hydrocarbons, carbon monoxide shift and carbon dioxide removal is therefore essential in conjunction with hydrogasification systems and will be discussed later under a separate heading (Section 7.4.).

The characteristics of the feedstock, as discussed in Chapter 4, are significant from several points of view. Aromatics are not generally converted under the operating conditions mentioned. This means that only non-aromatic hydrocarbons, including most of the paraffinic side chains of aromatic nuclei, are hydrogenated, whereas the rings themselves leave the gasifier as benzene, toluene and possibly xylene vapour. An important feature of all recycle hydrogenators, therefore, is an aromatics separator, which is a catch-pot for condensed aromatics obtained by quenching the outlet gas.

It has also been found that in the total absence of sulphur compounds the stainless steel lining of the reactor appears to have a catalytic effect on the thermal pyrolysis of hydrocarbon feedstocks that can assume catastrophic rates. A number of precautions are required to minimise carbon lay-down, among these the inclusion of 10 to 20 ppm of organic sulphur compounds in the hydrogenator feed, the addition of 8% steam and also the possibility of constructing reactor lining and interior of materials other than stainless steel, which appears to catalyse the cracking reaction and Boudouard equilibrium ($2CO \rightleftharpoons C + CO_2$), and must therefore be 'passivated'.

The 'empty vessel' gas recycle reactor has been used commercially for the gasification of light and heavy naphthas with end points of $115\,^{\circ}C$ and $180\,^{\circ}C$ respectively. Under experimental conditions substantially higher boiling feedstocks, *i.e.* jet fuel, kerosine and light gas oil with end points of up to 300 to $350\,^{\circ}C$ have been gasified. However, by-product yields gradually increase, typical production rates being of the order of $5 \cdot 5\%$, on a thermochemical basis, for light naphtha, $8 \cdot 5\%$ for heavy naphtha and 15% for kerosine. A further difficulty of processing higher boiling material is the near impossibility of completely vaporising the feed, especially at the higher pressures required for the gasification of higher boiling liquids. If these are still in the liquid state at the injection temperature, ($450\,^{\circ}C$), extensive cracking will occur. To avoid this, the hydrogen is sometimes heated separately to a much higher temperature than that of the liquid feed.

While it is possible to hydrogasify light, volatile hydrocarbons in an empty vessel of the Gas Recycle Hydrogenator type, the hydrogenation of heavier feedstocks is carried out more effectively in fluidised bed reactors of the type shown in Fig. 7.2 which is a diagrammatic representation of another process unit developed at the Solihull Research Station of the British Gas Corporation and is generally referred to as the Fluidised-bed Hydrogenator (FBH).

The Fluidised-bed Hydrogenator is again a heat-insulated high pressure cylindrical stainless steel vessel this time with separate inlets for fluidising gas, hydrogenating gas (which also acts as dispersant or atomiser) and oil. All are located at the bottom of the vessel. Product gas is withdrawn from a disengagement space near the top of the reactor. Circulation of the fluid bed is assisted by an inner riser tube through which solids and gas rise to the top of the bed. At the top, gas and solids are separated by a series of baffles and the solid particles are returned to the bed. A lower rate of gas flow in the annulus round the inner tube results in a net downward flow of solids in that region of the bed. Particles therefore descend as far as the gas distributor plate, where they are entrained upwards again by the more rapid flow in the inner core of the bed. Agglomeration and bogging of the bed is therefore prevented.

The fluid bed consists of coke particles, 0·1 to 0·4 mm in diameter, into which atomised, though not vaporised, heavy oil is injected. Since some carbon deposition will invariably occur under these conditions, provision must be made for the continuous withdrawal of some of the coke and, generally, for size reduction of particles that have grown excessively owing to carbon build-up. Coke fines are continuously returned to the bed.

The temperature in the bed, much as in the empty vessel hydrogenator, is of the order of 750 °C and both hydrogen and oil feed are preheated to at least 350 °C, the former sometimes to a higher temperature. The pressure in the fluidised bed hydrogenator is within the range of that used with empty vessel hydrogenators (10–100 atm) but in practice rather higher than the average values (50–70 atm v. 25–40 atm) which apply to commercial GRH plants.

The hydrogenating gas should again contain a high percentage (ca. 95 % vol) of hydrogen; steam can be added but is not absolutely essential since provision for carbon deposition on the coke particles

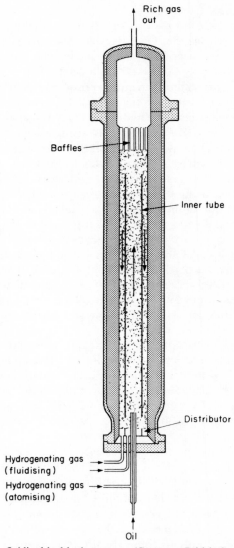

FIG. 7.2   The fluidised-bed hydrogenator. (Courtesy: British Gas Corporation.)

eliminates the need for protection against carbon formation. Sulphur compounds in either hydrogenating gas or feedstock are of no consequence in the process since conversion is entirely non-catalytic. However, the exit gas must be purified and the cost of H_2S removal may at times impose certain limits on the amount of sulphur that can be accepted.

It has been mentioned that aromatic compounds in the feedstock are not gasified under standard hydrogenator operating conditions. Since the concentration of ring compounds in heavier feedstocks is generally higher than that in light distillates, a large proportion of the feedstock thermal content will find its way into aromatic liquid by-products, which have to be removed from the reactor exit gas. In the case of heavy feedstocks some condensed ring compounds (naphthalene, anthracene) will be found in the condensate in addition to benzene and toluene, and the total yield of by-products in terms of heat content will be between 20 and 25 per cent.

Feedstocks that would be processed in fluidised-bed hydrogenators range from heavy gas oil to light or heavy crude oil and even fuel oils, although the latter will tend to yield much smaller proportions of clean gas than does gas oil.

The hydrogen gasification route is thus applicable to a wide range of feedstocks. However, different types of equipment are used and there are obviously also differences in process operating conditions.

7.3 HYDROGASIFICATION PROCESSES

7.3.1 Gas Recycle Hydrogenation for SNG Manufacture[1,3,8,10]

Typical feedstock characteristics, gas compositions, operating conditions and yields are shown in Table 7.1 for a number of hydrocarbons that have been processed in gas recycle hydro-genators.[2]

It will be noted that none of the gases obtained from the GRH reactor is acceptable as a substitute for natural gas because neither the Wobbe Index nor the Weaver flame speed factor is within the range required for interchangeability with standard natural gases (see Chapter 3). Direct manufacture of SNG by simple hydrogenation of a petroleum feedstock is therefore not practicable and some form of additional treatment becomes necessary. The aim of such a further treatment is firstly to reduce hydrogen content in order to lower the

TABLE 7.1
Hydrogenation of Distillates in Gas Recycle Hydrogenator

	Naphthas			Kerosine	Gas oil
Feedstock					
Rel. density, 15°C					
(water = 1).	0·72	0·72	0·69	0·78	0·84
Final boiling pt., °C	185	185	121	232	335
Carbon/hydrogen ratio	5·8	5·7	5·4	6·2	6·5
Aromatic content, % vol.	6·5	7·2	3·3	15·0	28·0
Operating conditions	—				
Pressure, atm	30·6	30·6	22·1	20·8	30·6
Preheat temperature, °C	425	403	406	462	Oil 382/Gas 654
Operating temperature, °C	721	727	724	749	741
Oil/gas ratio, g/m³	851	686	1 035	483	508
Hydrogenating gas, % vol					
CO_2	0·8	0·3	0·2	7·5	0·3
CO	3·7	3·7	3·4	3·1	3·1
H_2	92·4	94·2	95·9	86·6	89·8
CH_4	2·4	1·3	0·5	2·0	5·9
N_2	0·7	0·5	0·0	0·8	0·9
Product gas, % vol					
CO_2	0·7	0·7	0·2	6·8	0·2
C_xH_y (unsats)	2·4	1·1	3·6	1·0	1·6
CO	3·3	3·8	3·2	3·1	1·3
H_2	23·8	34·8	25·8	41·7	42·2
CH_4	44·0	36·8	42·9	31·3	36·0
C_2H_6	25·1	22·1	24·3	15·3	17·6
N_2	0·7	0·7	0·0	0·8	1·1
Calorific value, kcal/Nm³	9 797	8 718	9 854	7 257	8 049
Wobbe Index (metric)	11 381	10 926	11 657	9 300	14 690
Weaver flame speed	21·5	24·5	21·8	26·0	26·2
Thermal yield, %					
Gas	84·9	89·7	88·5	82·5	74·8
Liquids	12·1	8·3	8·5	13·5	21·0

flame speed, secondly to eliminate or reduce the ethane content, which ultimately could result in too high a calorific value, and thirdly as a result of both these exercises, to increase the methane content of the gas. There are a number of alternative possibilities to produce these effects and it is proposed to deal with them in turn.

If the hydrogenating gas supplied to a gas recycle hydrogenator has a sufficiently high carbon oxides content, and if these carbon oxides simply pass through the hydrogenator without reacting, one can, for example, reduce hydrogen content and at the same time increase the methane content by a subsequent methanation step.[3] Whilst this is technically feasible it presents problems: a considerable volume of carbon oxides has to pass through the GRH plant, where it acts purely as ballast; methanation of such high concentrations of hydrogen is extremely exothermic and would have to be carried out in a number of stages, and the presence of higher hydrocarbons in the gas raises its calorific value to an uneconomically high level. Finally, the economics of the process are against simple methanation because of the high cost of hydrogen and the unsatisfactory utilisation of the expensive gas in a simple methanation process.

As an alternative to the direct methanation by residual hydrogen of excess carbon oxides introduced into the system, one can also submit the effluent gas from a GRH or FBH plant to low temperature reforming. This results in the formation of methane and smaller quantities of carbon oxides and hydrogen from the ethane component of the gas, the remaining components being unaffected or reacting only to a small extent under the conditions of low temperature steam reforming. The effluent gas from the LT reformer now contains a suitable concentration of carbon oxides which can be reduced by means of residual hydrogen over a methanation catalyst. A flow sheet for the production of SNG by this same series of processing steps has been outlined.[10]

The advantage of the LT reforming/methanation route is that it removes excess higher hydrocarbons in the gas and thus reduces its final calorific value. On the other hand it still presents the problem of a high hydrogen concentration in the feed gas to the methanator and the consequent high degree of exothermicity of the methanation reaction.

A final route that will be discussed in greater detail in the following section is the use of the gas recycle hydrogenator effluent gas in the catalytic hydrogasification of a light hydrocarbon such as naphtha or LPG. This is in fact a re-equilibration of the gas components, methane, ethane, hydrogen and carbon oxides, in the presence of additional light hydrocarbon feedstock. Since the steam reforming of the additional feedstock can be slightly endothermic, it effectively absorbs the process heat generated by methanation. It also utilises residual hydrogen more effectively and seems the most likely process route to SNG where hydrogenators are used as a first processing step.

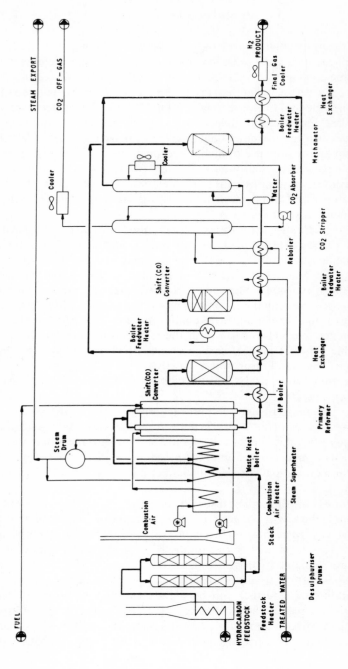

FIG. 7.3 Flow diagram showing application of ICI continuous reformer to hydrogen production. (Courtesy: Humphreys and Glasgow Ltd.)

7.3.2 Catalytic Hydrogasification System

The direct reaction between hydrogen and various hydrocarbons can be carried out under both catalytic and non-catalytic conditions. The non-catalytic reactions that we have discussed so far take place on the surface of steel vessels, and in the fluid coke beds under similar conditions, and could therefore be due to a certain catalytic element; however, no catalyst was specifically designed to promote them and the only catalytic hydrogasification in the true sense of the word is the one considered here which is used generally in conjunction with a first low temperature reforming stage.

It will be remembered that low temperature reforming produces a gas that consists of about 60 % vol of methane, 20 % vol of hydrogen and 20 % vol of carbon oxides, the latter mainly in the form of carbon dioxide. This product is obtained under conditions described in Chapter 6 and clearly requires further processing before distribution in the form of SNG. A hydrogasification step that simultaneously mops up surplus hydrogen and involves the further processing of additional feedstock and also results in a better utilisation of the steam of the first stage is clearly a most attractive method of improving both overall economics of the process and inter-changeability of the product gas (Fig. 6.3).

The reactions in a hydrogasifier take place at a temperature somewhat lower than that of the low temperature CRG or similar reactor. The catalyst used in a hydrogasifier is usually the same as that of the LT reformer, *i.e.* a high nickel material on an alumina base; however, its operation differs from that of the low temperature steam reformer in that the catalyst can be, and usually has to be, regenerated after a period of use. This has the double effect of, on the one hand decreasing the importance and the danger of sulphur contamination and, on the other hand promoting the removal of deposited polymer carbon from the catalyst. Catalyst regeneration is usually effected by means of hydrogen, *i.e.* instead of passing a mixture of low temperature effluent gas, steam and additional hydrocarbon vapour, the catalyst is only reduced with hydrogen of a reasonably high degree of purity. The reactions that take place in a catalytic hydro-gasification reactor are extremely complex. Not only does a high degree of methanation reduce the concentration of both carbon oxides and hydrogen, but in addition residual steam in the low temperature reformer effluent reacts with hydrocarbon vapour admitted into the second reactor. The lower reaction temperature results in an equilibrium mixture far closer to interchangeable SNG than that in single stage LT reforming.

Table 7.2 compares composition and combustion characteristics of hydrogasified naphtha after one stage of reforming, after reforming and hydrogasification and finally after a methanation step designed to remove any residual traces of hydrogen in the gas.

TABLE 7.2
LT Reforming—Hydrogasification
Gas Composition and Combustion Characteristics

	SNG delivery pressure		34 atm
	Naphtha feedstock C/H ratio		5·9:1 by weight
	Overall steam/naphtha ratio		1·2:1 by weight
	LT reformer	Hydrogasifier	Methanator
Temperature, °C			
inlet	450	400	280
outlet	515	488	340
Gas composition, % vol (dry)			
CH_4	65·52	70·93	77·16
H_2	11·75	6·54	0·63
CO	0·87	0·67	0·07
CO_2	21·86	21·86	22·14
Gross heating value, kcal/Nm³	6 904	7 309	7 737
Relative density, 15°C (air = 1)	0·7041	0·7372	0·7659
Wobbe Index	8 228	8 513	8 841
Weaver flame speed factor	24·3	20·5	14·7

The economic advantage of hydrogasification over other forms of processing naphtha to SNG lies essentially in its better steam utilisation. It is normal to dilute naphtha vapour in a low temperature steam reformer with about 2 kilograms of steam per kilogram of naphtha. A combination of low temperature reforming and catalytic hydrogasification permits this ratio to be reduced to about 1:1, mainly because of the additional naphtha vapour injection into the second stage. It can be calculated that this reduction of steam consumption in the process corresponds to an improvement of about 4% in the thermal efficiency of the conversion.

In addition, the combined low temperature reforming-hydrogasification route is also advantageous in regard to investment. If one assumes, for example, the first stage, low temperature reforming, is the same for two units, one using methanation, the other hydrogasification, then the LTR/HG combination processes between 1·8 and 2 times the amount of feedstock, owing to the additional intake into the second stage. The specific investment, with reference to feed conversion or SNG output, is thus substantially reduced.

Hydrogasification, instead of direct methanation of the gas as it emerges from the low temperature steam reformer, has been adopted by two of the three commercial low temperature reforming systems, the Lurgi Gasynthan and the British Gas Corporation's CRG processes; no corresponding route, as far as could be ascertained, was available in conjunction with the Japan Gasoline Co.'s equivalent process.

7.3.3 The Hydrogenation of Heavy Fuels[5,8]

Although the hydrogasification reactions that result in the breakdown of the larger hydrocarbon molecules are exothermic, the reactions themselves cannot be initiated at temperatures much below 700 °C even if hydrogen pressure is raised to 50 atm or more. Rapid preheating of hydrocarbon feedstocks to this temperature results in relatively little carbon formation in the case of light feedstocks such as naphtha or LPG but the competing pyrolysis reaction plays a much more important role in the case of heavy fuels.

Various approaches can be used to minimise cracking and carbon formation, such as preheating hydrogen and hydrocarbon feeds separately, the former to a higher temperature than the latter, but some pyrolytic carbon formation will inevitably occur with the higher boiling feedstocks. It is to deal with this situation that the fluid-bed hydrogenator, which has been discussed previously (Section 7.2), was developed.

The fluid-bed hydrogenator has been used so far only on an experimental basis, although a demonstration plant at Osaka, Japan, has been operating for several years to gasify crude oils of different gravities and even residual fuel oils. Feedstock characteristics, operating conditions, composition of the hydrogenating and product gases and heat balances for the conversion process for a number of conversion runs are summarised in Table 7.3. As with the GRH process, gas combustion characteristics are not fully acceptable and methanation of the product gas is essential.

It will be gathered that the gasification of crude oils and residual fuels in a fluid-bed hydrogenator, much as that of lighter feedstocks in

TABLE 7.3
Fluidised-bed Hydrogenator

	Crude oil		
Feedstock			
Relative density, 15°C (air = 1)	0·86	0·89	0·92
Conradson carbon, % wt	3·5	3·9	5·7
Operating conditions			
Pressure, atm	50·0	50·0	50·0
Preheat temperature, °C	Oil 350/Gas 600	Oil 350/Gas 625	Oil 350/Gas 650
Operating temperature, °C	750	750	750
Oil/gas ratio, g/m³	624·8	598·1	505·2
Hydrogenating gas, % vol			
CO_2	0·3	0·2	0·3
CO	3·1	3·1	3·1
H_2	94·0	94·1	93·9
CH_4	1·7	1·6	1·8
N_2	0·9	1·0	0·9
Product gas, % vol			
CO_2	0·2	0·2	0·2
C_nH_m (unsats)	0·3	0·2	0·2
H_2	36·3	35·9	36·5
CO	2·9	2·6	2·8
CH_4	45·9	46·6	46·0
C_2H_6	13·5	13·6	13·4
N_2	0·9	0·9	0·9
Calorific value, kcal/Nm³	7 840	7 829	7 820
Wobbe Index, kcal/Nm³	10 760	10 797	10 803
Weaver flame speed	25·9	25·4	26·1
Heat balance			
Gas	72·9	71·5	70·1
Liquids	21·7	22·4	23·2
Carbon deposited	1·3	1·6	2·1

a gas recycle hydrogenator, produces, in addition to gas, substantial quantities of aromatic liquids and some carbon, irrespective of differential gas/oil preheating, a high gas/oil ratio, a high hydrogen pressure and an operating temperature not exceeding 750°C.

The liquid products formed in the FBH reactor when processing crude oil represent a rather higher proportion of the total than in the case of the GRH process (21 to 23 versus 3 to 8 per cent). Similarly, the content of higher aromatics such as naphthalene and anthracene is substantial, whereas these are practically absent in naphtha feedstocks boiling up to 115°C and occur only in traces if naphtha boiling up to 120 °C is gasified. Both the lower gasification yield and the shift towards relatively useless products—at the expense of benzene and toluene—are thus features of the hydrogasification of heavier feedstocks.

The choice of a light feedstock and of the technically simpler empty vessel GRH reactor rather than the gasification of heavy feeds in an FBH plant will thus be dictated by relative feed economics and, of course, feedstock availability. Provided low gravity feedstocks are available and their price is not much higher than that of crude oil, and both these provisos must be considered doubtful in many locations at the moment, hydrogasification in a GRH plant will clearly be the preferred solution. If, on the other hand, neither naphtha nor kerosine is obtainable or if their cost is much higher than that of crude oil, FBH processing must be considered. This topic will, however, be resumed in greater detail in Chapter 11 under the heading of SNG economics.

7.4   THE MANUFACTURE OF HYDROGEN

The conversion of heavier feedstocks, the removal of contaminants from feedstocks destined for catalytic conversion and the elimination of carbon monoxide from a gasified feedstock by conversion into methane, all require hydrogen. In certain instances the hydrogen for these reactions can be supplied in part or in total from the reaction products of a prior gasification reaction, *e.g.* in the methanation of low temperature reformer gas, but in most instances additional hydrogen or even the entire hydrogen consumption is met by manufacturing the gas separately in plant specially designed for the purpose.

The manufacture of hydrogen is not altogether different from that of SNG. In fact, most gasification processes can be biased to produce hydrogen, generally by operation at a high temperature when the equilibrium of the steam reforming reaction, for example

$$CH_4 + H_2O \rightleftharpoons CO + 3H_2$$

tends to shift to the right. It can therefore be said quite generally that gasifying hydrocarbons at higher temperatures will produce hydrogen, and furthermore, that the presence of excess steam will be beneficial. Finally, since the reforming, reaction is endothermic it will be necessary to supply heat to produce hydrogen (while its disappearance through carbon oxide-methanation reactions will be accompanied by heat evolution).

Feedstocks for hydrogen manufacture again range over the whole gamut of hydrocarbons from methane to coke. Much as in SNG conversions, yields from the lighter feedstocks are higher, investment is lower and catalysts can be used where the feedstock is either sulphur-free or can be desulphurised to a very low sulphur content.

Gasification to produce hydrogen can be by pyrolysis, which is inefficient, *i.e.* produces polymers, tars, carbon, coke and other by-products in large quantities, by hydrolysis and oxygenolysis, the endo- and exothermic routes to gaseous products. Clearly, in the light of their contradictory thermal requirements, the two are best combined, *i.e.* the feedstock is reacted with steam and oxygen, as in the Shell, Texaco and other similar so-called partial oxidation process routes.[4] Alternatively, the heat of reaction for hydrolysis can be supplied by outside firing of catalyst-filled reactor tubes, *e.g.* in the ICI, Selas, Hercules and similar conversions. Finally, heating of the reaction system can be achieved by cycling hot gases and reactants alternately through the same reactor. However, this is rarely commercially feasible since such systems cannot be pressurised, and hydrogen is invariably required under pressure.

The most usual methods of producing hydrogen are thus the catalytic high temperature steam reforming of a desulphurised light

TABLE 7.4
Hydrogen Manufacture

	Shell Process	ICI Process
Feedstock range	any hydrocarbon	gas, LPG, naphtha
typical	fuel oil	naphtha
Preheat temperature, °C		
Feedstock	236	300
Steam	236	400
Oxygen	246	—
Exit temperature, °C	1300–1400	815
Operating pressure, atm	32	12
Gas composition, % vol (ex. H_2O)		
CO	46·9	14·0
CO_2	4·3	13·6
H_2	46·2	70·8
CH_4	0·3	1·6
N_2	1·4	—
H_2S	0·9	—
Gasification yield, % vol	85	81

hydrocarbon feedstock in a catalyst-packed tubular reactor; and the steam/oxygen reaction of heavier feedstocks in an empty vessel, at an even higher temperature.

Table 7.4 lists operating conditions for two typical conversions of the type described, the Shell gasification, *i.e.* partial oxidation, process for fuel oil and the ICI continuous high temperature reforming process for naphtha.

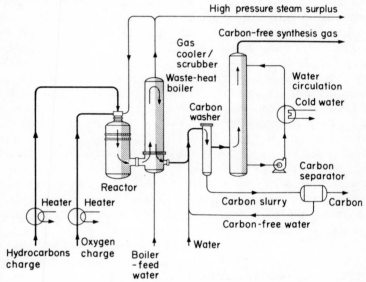

FIG. 7.4   Diagram of the Shell partial oxidation process. (Courtesy: Bataafse Internationale Petroleum Maatschappij NV.)

Neither gas can be used directly in a hydrogasification process and both have to be cooled and treated in order to arrive at a good quality hydrogenating gas. Flow diagrams for the two processes are shown in Figs. 7.3 (ICI process) and 7.4 (Shell process) with the necessary purification and conversion facilities included in the ICI process diagram.

Oxygen gasification processes such as the Shell or Texaco processes can convert any liquid petroleum hydrocarbon into a gas of approximately the composition listed in Table 7.4. However, with some modifications, this type of gasification can also be applied to hydrogen manufacture from solid fuels. Both atmospheric and pressurised steam/oxygen solid fuel gasifiers have been developed. Examples of the former are the Koppers-Totzek and Winkler plants, both developed during the war in Germany and based on some form of fine dispersal or fluidisation of pulverised coal or semi-coke. While they are interesting early developments, they suffer from inherent disadvantages such as small output per unit, a tendency to incomplete gasification and loss of fuel in the fly ash in the Winkler process and in the molten slag in the Koppers-Totzek process. Inability to gasify other than strictly defined grades of fuel are additional drawbacks.

Attempts have recently been made to modernise solid fuel gasification, and the second generation processes, mostly of American origin, are discussed in Chapter 9 which deals with the manufacture of SNG from coal. In the present context, one would only like to draw attention to the fact that the US Bureau of Mines process, the IGT HYGAS process (in one of its versions), the BI-GAS process and the older Lurgi process generate a gas under pressure by reacting coal with steam and oxygen.

In the light of the high cost of oxygen, which results in extremely expensive hydrogen if made by any of the above routes, a number of attempts have been made to substitute air as an oxidant or to use some other source of external heat in liquid or solid feed hydrogen plants. Much technical ingenuity has been used and some of the routes are commercially feasible.

If air rather than oxygen is used some way must be found to separate combustion gases, including carbon oxides and hydrogen, from the accompanying nitrogen.

The previously mentioned fixed bed cyclic processes achieve this by a time separation—air blowing and gas making phase alternate—but they are difficult to operate under pressure. In more modern versions the fuel is circulated, being air oxidised and heated in one reactor and steam reformed and cooled in another, with streams of hot and cold particles circulating between the reactors. The Union Carbide process, a version of the IGT, the Exxon FLEXICOKE system and others are based on these principles.

A final route to hydrogen uses steam without oxygen and again ways and means have been found to introduce heat into a pressurised system by circulating an inert heat carrier, slag or a molten salt (Agglomerated Ash and Kellogg processes) or by allowing an exothermic reaction to take place in the reactor bed, as in the CO_2-acceptor process. All these technologies are reviewed in Chapter 9.

The purification of hydrogen produced in any of the above mentioned types of generator proceeds along similar lines. The following steps are involved, but depending on feedstock and process used, some of them may be by-passed:

Gas cooling
Removal of soot, fly ash, carbon black
Shift conversion: $CO + H_2O \rightarrow CO_2 + H_2$
Gas cooling
Removal of sulphur compounds (H_2S, COS)*
Removal of CO_2

Little need be said about cooling, except that it is used to generate steam, that specially designed boilers must be used if carbon black has been formed which might clog narrow passages, and that for that reason direct water injection is sometimes used rather than indirect heat exchange.

The subsequent removal of solids from the gas uses water injection and cyclone separators; usually it is desirable to recycle soot, tar and carbon black to the gasifier and the form in which they deposit is therefore important. It is clear, however, that in many instances no carbon removal will be needed; light hydrocarbons are normally gasified without carbon formation, and in catalytic processes particularly, there is less tendency towards pyrolysis.

The shift conversion that follows is essential if pure hydrogen rather than a mixture of reducing gases is required. Having made sure that surplus steam is available, the gases are passed over a cobalt catalyst

* It is sometimes preferable to remove sulphur before shift conversion.

TABLE 7.5
Hydrogen Gas Composition at Various Stages of Purification

Composition, dry % vol	Feedstock heavy fuel oil		ex H₂S removal	ex CO shift	ex CO₂ removal
	Gas ex Shell gasifier	ex carbon catcher	ex H_2S removal	ex CO shift	ex CO_2 removal
H_2	46·2	46·2	46·6	63·6	95·5
CO	46·9	46·9	47·3	0·3	0·45
CO_2	4·3	4·3	4·4	34·9	2·25
CH_4	0·3	0·3	0·3	0·2	0·3
N_2 + Ar	1·4	1·4	1·4	1·0	1·5
H_2S + COS	0·9	0·9	nil	nil	nil
Carbon	1 ppm	nil	nil	nil	nil

at a temperature of 250 to 300 °C, which results in a residual concentration of 0·5 % vol of carbon monoxide or less.

However, the gas now contains substantial amounts of acidic compounds, some hydrogen sulphide and between 20 and 40 % vol of carbon dioxide. Absorbents of the type discussed in Chapter 4, which remove acidic gases at low temperature and release them on boiling (amines, glycols, potassium carbonate, alkazid fluids, etc.), are used to purify the gas of sulphur compounds and CO_2 either in one joint operation or separately. Typical gas compositions produced in the various purification stages are listed in Table 7.5.

7.5   CONCLUSIONS

Hydrogasification processes are among the most versatile routes to SNG. They can be adapted to process a wide range of liquid fuels and the manufacture of the necessary hydrogen can be based on practically all liquid and solid fuels, opening up possibilities of combining the conversion of several feedstocks in order to arrive at the most economic route.

It is, however, difficult technically to produce hydrogen without the use of oxygen and, the latter being always expensive, methods are being developed particularly in connection with coal gasification, to eliminate the need for an oxygen plant. Some of these new technologies will be discussed in more detail in Chapter 9.

Hydrogen gases, once they are produced, by whichever route, must generally be purified for use in hydrogasification processes, and standardised processes are available for the several steps involved. A pure hydrogen gas can then be used to convert petroleum fuels into SNG or at least into intermediate products that can be processed to become interchangeable with natural gas.

REFERENCES

1. Davies, H. S., Lacey, J. A. and Thompson, B. H. (1969). Processes for the manufacture of natural gas substitutes, *J.I.G.E.*, **9**, 375.
2. Dent, F. J., Edge, R. F., Hebden, D., Wood, F. C. and Yarwood, T. A. (1956–7). Experiments on the hydrogenation of oils to gaseous hydrocarbons, *Trans. Inst. Gas Engrs*, 106, p. 594. UK Gas Council Res. Comm. GC 37 (1956).

3. Dent, F. J. and Hebden, D. (1968). *The Gasification of Oil to Yield High Calorific Value Gases*, World Power Conference, Moscow, Section A2, p. 207.
4. Kuhre, C. J. and Sykes, J. A. (1973). *The Shell Gasification Process for the Substitute Natural Gas Industry*, Inst. Gas Tech., SNG Symposium I.
5. McMahon, J. F. (1973). *The FBH Process—SNG Production from Crude Oil*, Paper 13, Inst. Gas Tech., SNG Symposium I.
6. Murthy, P. S. and Edge, R. F. (1962). The hydrogenation of oils to gaseous hydrocarbons, *J.I.G.E.*, **3**, 459.
7. Binay, B. *et al.* (1966). Production of high Btu gas from light distillate by continuous pressure hydrogenation, *Industr. Engng Chem. Process Design and Development*, **5**, 247.
8. Thompson, B. H. and Conway, H. L. (1972). *The Hydrogenation of Hydrocarbons in Relation to the Manufacture of Substitute Natural Gas*, Paper to 65th Annual Meeting of the American Institute of Chemical Engineers.
9. Thompson, B. H., Majumdar, B. B. and Conway, H. L. (1966). The hydrogenation of oils to gaseous hydrocarbons, *J.I.G.E.*, **6**, 415.
10. Tittle, R. W. and Hartley, W. (1973). *The Potential of the Gas Recycle Hydrogenator in the Production of SNG*, Paper 12, Inst. Gas Tech., SNG Symposium I.

Chapter 8

The Gasification of Crude Petroleum

8.1 INTRODUCTION

The conversion of heavy petroleum fuels into SNG can proceed by a number of different routes. One can, for instance, as discussed in some detail in the previous chapter, hydrogenate crude oil or even heavy residues in a fluidised bed and produce gases that, by some relatively minor further processing, can be made interchangeable with natural gas. The process obviously requires hydrogen manufacture as a first step, and this itself may need oxygen as a processing intermediary; the result is a fairly complex chain of conversions that require close integration, an overall conversion system of a high degree of technical complexity that cannot be constructed cheaply.

The oxygen gasification of heavy fuels, which is one of the ways of making hydrogen, can also be used as the basic SNG conversion step; in other words the entire feedstock can be gasified to produce a low calorific value gas (synthesis gas) which can, in turn, be used to produce methane. While this route involves a relatively simple series of processing steps, it is not very efficient thermally, the heat generated in the methanation process being too little for utilisation in other stages such as power generation for air separation and oxygen manufacture.

A third approach to the conversion, particularly of crude oils with their complex make-up of hydrocarbons ranging from very light gases to heavy fuel oil, is the combination of crude oil refining with gasification. It has the advantage, if carried out on a sufficiently large scale, of either treating the whole crude or of using the most appropriate gasification route for each particular fraction of the crude. It also permits the simultaneous manufacture of low sulphur liquid fuels, desulphurised with the aid of the hydrogen gas that has to be produced at some stage, in cases where production of such fuels is

136

justified by the demand or by technical–economic considerations of the gasification process.

The routes to SNG that are reviewed in this chapter are thus all basically suitable for heavy distillates, crude oils and fuel oils. They do not include the Fluid-Bed Hydrogenator, which has already been discussed as a processing tool for SNG manufacture from crude oil in Chapter 7. However, various hydrocracking process routes; FLEXICOKING—a thermal cracking route including coke gasification; conversion of heavy fuel oil by partial oxidation with oxygen and the alternatives of complete conversion into SNG, or the simultaneous production of SNG and low sulphur fuel oil are reviewed in some detail.

8.2 COMPLETE GASIFICATION OF CRUDE OIL[12,13]

Under this heading it is proposed to deal essentially with situations in which one aims at converting the bulk of the liquid feed into SNG. Only a minimum of by-products such as aromatics, coke or asphalt are to be produced and there is no intention to co-produce low sulphur fuel oil. Our definition thus respectively eliminates from consideration both the FBH process,[7] discussed in Chapter 7, and oil refinery processes yielding both SNG and liquid fuels, which are discussed in Section 8.3 of this chapter. However, it still leaves the other three main gasification processes, hydrocracking, partial oxidation and coking, for the principal conversion step. It also allows for two alternative ways of processing namely gasification of the complete feedstock, as a first step, or prior fractionation into more uniform and therefore more manageable hydrocarbon fractions, followed by gasification.

8.2.1 Hydrocracking–Hydrogasification[2]
Processing schemes for the conversion of crude oil that have been developed from existing petroleum refining systems include the various so-called hydrocracking routes. Hydrocracking is an adiabatic conversion in which preheated feedstock and recycle gas containing hydrogen are charged to a pressurised reactor containing an appropriate catalyst. It can be applied to a wide-boiling material such as whole crude, but it is more common to fractionate the feedstock first and to treat the individual fractions separately,

subjecting only distillates to the actual hydrocracking step, *i.e.* gasifying lighter materials by steam reforming, heavier products by other processes to be described later.

The purpose of hydrocracking, it should be clearly understood, is not the gasification of the feedstock, either crude or distillate, but its preparation for further treatment by appropriate gasification steps. Its consumption of hydrogen is therefore much lower than that of the FBH process. The attraction of fractionation prior to hydrocracking is, amongst others, the possibility to combine light gasification feedstocks obtained by crude fractionation with additional material prepared by hydrocracking and to submit the combined stream to low temperature reforming or a similar gasification process route.

Hydrocracking is a term usually applied to the treatment of a hydrocarbon with hydrogen at a temperature of 340–420 °C, a pressure of 80 to 150 atm and an oil/gas ratio (gas recycle) of 1·5 to 2·0 Nm^3 per litre. The main effect of this pretreatment of either the whole crude or of the gas oil fraction is a substantial reduction in carbon-forming components, and it has in fact been found that the required degree of hydrocracking can be assessed by testing for the residual Conradson Carbon content, a concept discussed in Chapter 4. In addition, the products are sulphur-free or low in sulphur, most of the latter being converted into H$_2$S.

A number of commercial hydrocracking processes are available for both distillate and residual feedstocks. Several of these have been proposed as intermediate gasification stages. They include the following three fixed-bed catalytic routes:

—Isomax   developed by UOP and Chevron Oil Co.[4]
—Unicracking   due to Union Oil Co. of California
—JHC   developed by Exxon Research and Engineering
(Jersey Hydro-
Cracking)

and a moving catalyst system:

—H-Oil   developed by Hydrocarbon Research Inc.

Typical characteristics of the heavy feedstock and yields of the light products, with emphasis on those properties that are relevant to gas making, are shown in Table 8.1.

All residuum hydrocracking processes have in common the problem of metal deposition on the catalyst. Since all crude oils and

TABLE 8.1
Hydrocracking Yields and Product Quality

Feed	Straight-run gas oil	Cat. cracker cycle oil
Sp gr, 15/15 °C	0·914	0·94
ASTM distillation rage °C	350–565	300–430
Sulphur, % wt	2·4	1·4
Nitrogen, ppm	750	430

Operation	Straight-run gas oil			Cat. cracker cycle oil	
	Gasoline	Gasoline + jet fuel	Gasoline + middle distillate	Gasoline + middle distillate	Gasoline
Yields, vol/100 vol feed					
C_2	1·0	1·0	1·0	1·8	5·3
C_3	3·8	3·3	3·0	2·6	9·2
Isobutane	13·3	8·8	6·9	1·9	6·1
n-Butane	3·8	3·6	2·4		
Light gasoline	32·2	23·8	16·6	48·9	33·9
Heavy gasoline	74·7	44·1	43·9	65·9	81·8
Middle distillate		43·0	48·6		
H_2 consumption, Nm^3/l	0·35	0·34	0·32	0·43	0·57
Product inspection, Light gasoline					
Boiling range, °C	C_5–80	C_5–80	C_5–80	C_5–170	C_5–80
Sp gr, 15/15 °C	0·657	0·657	0·657	0·747	0·669
Paraffins/naphthenes/aromatics % vol	82/15/3	82/15/3	82/15/3	37/54/9	63/36/1
Product inspection, Heavy gasoline					
Boiling range, °C	80–204	80–160	80–177		80–200
Sp gr, 15/15 °C	0·759	0·747	0·751		0·763
Paraffins/naphthenes/aromatics % vol	48/47/5	46/49/5	47/50/3		40/54/6
Product inspection, Kerosine					
Boiling range, °C		160–290	177–327	170–350	
Sp gr, 15/15 °C		0·802	0·816	0·846	

fuel oils contain greater or lesser concentrations of ash-forming metal compounds, *e.g.* salts or organic complexes of sodium, calcium, iron, nickel, vanadium and other metals, they cannot be converted into lighter petroleum fractions or gases without, at the same time, forming hydrocarbon-insoluble metal salts. If the light products are subsequently vaporised the metal salts will be deposited on both catalyst and metal surfaces.

All catalytic hydrocracking systems, under those circumstances, tend to be deactivated in the course of operation, particularly so by vanadium-containing feedstocks; catalysts have to be regenerated or replaced either periodically in fixed-bed processes, or continuously in moving- or fluid-bed systems. Deactivation is, as a rule, much less of a problem when processing distillates.

The design of hydrocracking reactors is complicated by the need to remove reaction heat by one of the following mechanisms:

—injection of cold recycle gas;
—injection of cold recycle liquid;
—interstage heat exchange and cooling.

Fairly complex designs involving both external and internal recycling and intensive heat exchange between entering and issuing streams have therefore been developed.

Product yields and properties vary according to severity of treatment; clearly the higher the temperature, pressure and gas/oil ratio and the longer the residence time the greater the degree of conversion of heavier into lighter molecules. On the other hand, severity of cracking is limited, particularly for fixed-bed processes, by the possibility of carbon deposition in the catalyst bed. This is less of a problem in a moving bed, such as that used in the H-Oil process where constant attrition tends to remove the deposits.

Catalysts used in the Unicracking and JHC processes are typically platinum and palladium on a zeolite base, and they are claimed to be extremely resistant to sulphur, nitrogen and other contaminants in the feedstock. Other sulphur-resistant catalysts are based on tungsten, cobalt and molybdenum oxides.

If it is desired to convert the bulk of the whole crude oil into SNG with the aid of hydrocracking a number of alternative routes can be envisaged. The first of the schemes shown in Fig. 8.1 illustrates a crude oil-based gas plant consisting of crude oil topping and vacuum distillation of the residue; hydrofining and low temperature steam

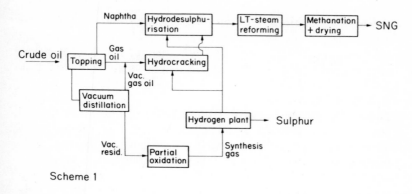

Scheme 1

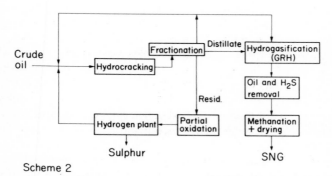

Scheme 2

FIG. 8.1 Complete gasification schemes based on hydrocracking.

reforming of desulphurised light fractions; hydrocracking of gas oils produced in both atmospheric and vacuum distillation steps, partial oxidation of the vacuum residue and conversion of the synthesis gas produced into hydrogen needed for the hydrofining and hydrocracking units.

Alternatively, as shown in the second of the schemes in Fig. 8.1, a gas recycle (GRH) hydrogenator can be used as the main gasification unit. Under these circumstances, hydrocracking of the crude oil should precede the fractionation step, and an oil and H_2S removal step will be required to purify the exit gas from the distillate treating—GRH reactor before it is methanated, dried and sent out. Residue from the fractionation would, as before, be converted by partial oxidation into hydrogen needed for the GRH reactor.

Yields of gas in the two schemes will be similar, and given the same degree of heat integration, gas yields of 84 to 87 % of the heat content of the feedstock should be obtainable, between 13 and 16 % being lost in the form of aromatics, sulphur, carbon and escaping process heat.

8.2.2 Partial Oxidation[10]

The partial oxidation of hydrocarbon materials ranging from natural gas to coal by means of oxygen has been described in Chapter 7 as a means of manufacturing hydrogen, which can then be used as a reagent in the further processing of liquid and solid fuels in the various steps that ultimately lead to SNG. Alternatively, a partial oxidation process can also be used to prepare a synthesis gas, consisting mainly of hydrogen, carbon monoxide and carbon dioxide, which can itself be treated to yield a methane-rich SNG.

The processing of the gaseous raw material follows exactly the steps discussed in the previous chapter. Oxygen is manufactured in a tributary air separation plant or purchased externally; oxygen, fuel and steam are injected under pressure into a ceramic-lined gasification reactor and the resultant gases are rapidly cooled. Again different cooling mechanisms, i.e. direct cooling by water injection or heat transfer in a specially designed boiler, can be used. And again the scrubbed cooled gas undergoes a carbon monoxide shift in a catalyst-filled reactor.

However, if one aims at the production of SNG, only part of the carbon monoxide is reacted with steam, the reason being that it is intended to finish up with almost pure methane. Therefore, shift conversion and subsequent CO_2 reduction must produce a gas in which the relative proportions of the components are aCO_2, bCO and $(4a + 2b)H_2$, and this implies that about half the effluent gas should by-pass the shift converter if the bulk of the CO_2 is to be removed in the next step.

Having thus prepared a 'synthesis' gas containing carbon monoxide and hydrogen roughly in the ratio of 1:2, we must be prepared for serious difficulties in the next step on the route to SNG, namely methanation. This reaction is discussed in some detail in Chapter 10, where it is shown that, owing to its high degree of exothermicity, ways and means must be found to control the reaction between CO and H_2 if the concentration of the gases is high. A number of cooling systems such as cold product gas recycle, reaction in stages with intermediate cooling, water injection and steam dilution have been proposed.

However, the methanation process is basically the same, whether it is used to remove the last traces of carbon oxides from hydrogen, to eliminate residual hydrogen from SNG or to convert pure synthesis gas into SNG. Extremely robust catalysts that are steam- and water-resistant, together with high rates of product recycle and interstage cooling, have to be used where SNG is prepared from gases containing high concentrations of carbon monoxide and hydrogen.

It should be borne in mind, however, that the temperature of methanation, about 300 °C, is insufficient by itself to generate high pressure superheated steam. This means that the electric power requirement of air separation cannot be met by simple heat exchange. And since the partial oxidation route, as described, requires a large volume of high pressure oxygen and, therefore, considerable amounts of electrical energy for air and oxygen compression, it becomes difficult, if not impossible, to design an efficient SNG plant on this basis.

8.2.3 Coking

It will be noted that the two gasification systems described so far as applicable to the gasification of heavy petroleum products have involved a partial oxidation stage and therefore require the availability of oxygen. The manufacture of oxygen by air separation on a scale commensurate with oil gasification plants of standard US capacity (120–250 million ft^3 per day) is very expensive in regard to both investment and operating costs. Energy in the form of electricity is required and unless the latter is extremely cheap, other approaches to gasification not involving electric power will generally be preferred.

A number of approaches to gasification of heavy products which do not involve the use of oxygen have been proposed and among these the so-called FLEXICOKING system developed by Exxon Research is one of the most promising.[5] A simplified flow sheet of a FLEXICOKER is shown in diagrammatic form in Fig. 8.2 and it is easy to see that such a system lends itself extremely well to the conversion of almost 100% of the thermal content of a crude oil feedstock into gaseous product.

In a FLEXICOKER, which is in fact a modification of the by now well established fluid coking system,[3] crude oil or residuum is fed into a fluid-bed reactor charged with hot coke particles. This results in extensive thermal cracking of the liquid feed with some carbon deposition on the circulating coke particles. The overhead volatilised

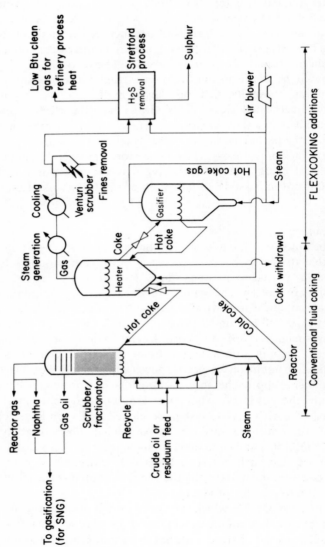

Fig. 8.2 FLEXICOKING—simplified flow plan.

products are fractionated in an integral column and naphtha and gas oil can be withdrawn as separate streams and gasified to produce SNG by the methods described in previous chapters. 'Heat for the endothermic coking reactions is supplied by circulating coke between the reactor and heater vessels. More than 95 % of the coke produced in the reactor is transferred *via* the heater to the gasifier vessel where it is gasified with air and steam. The remainder of the coke is withdrawn from the heater to purge feed ash and metals. Hot gas from the gasifier passes through the heater where it is cooled to provide a portion of the reactor heat requirement. The remainder of the heat is supplied by the coke circulating between the gasifier and heater. The gas leaving the heater is cooled via steam generation, scrubbed of fines and desulphurised using the Stretford Process for H_2S removal. The resulting coke gas product is a clean low Btu gas which can be used to supply refinery process heat requirements.'

It will be noted that under these circumstances a residual or crude oil feedstock can be converted completely into SNG with only a small amount of by-product coke being formed simultaneously. Instead of oxygen only air is used and the investment and operating costs of the plant are accordingly reduced.

A simpler version of a crude oil-to-SNG plant would not use the FLEXICOKING system but would simply fractionate the crude and coke the residual fraction, only the fluid coker overheads being gasified together with the crude oil light ends. Coke production under these circumstances would of course be much higher since none of the petroleum coke is gasified, and furthermore middle distillates that cannot be gasified by low temperature reforming methods would also be obtained as such.

If all the middle distillate formed in such a fluid coking plant is to be converted into SNG, *e.g.* by hydrogasification, additional hydrogen will be required—over and above that needed to desulphurise the low temperature reformer feed—and this can be produced from recycled reformer gas or refinery gas/LPG, or by partial oxidation of a heavy feedstock. The overall conversion in this simplified process thus becomes one of crude oil into SNG with surplus coke and distillate withdrawn from the fluid coking plant. However, the additional hydrogen requirement and the greater complexity of the gasification equipment more than outweigh the savings due to the elimination of the coke gasification section.

The FLEXICOKING route, which produces clean refinery fuel

from coke thereby freeing higher hydrogen fuels for SNG production and/or hydrogen manufacture, is believed to be both technically and economically superior to other ways of gasifying crude oil.

8.3 ENERGY REFINERIES[1,6]

The complete conversion of liquid fuels into SNG is still relatively high in its consumption of hydrogen, energy and sometimes oxygen. One of the means of reducing the cost of gasifying crude or residual fuel is to confine oneself to the gasification of the lighter fraction and to allow the heavier products, usually heavy gas oil or heating oil, to emerge as premium value by-products of the process. It is, however, then desirable to remove excess sulphur from these heavier materials.

The outcome of this type of conversion is therefore a relatively high proportion of SNG while most of the remainder of the feedstock is converted into low sulphur fuels in the heavy distillate range.

The reason why it is both economic and technically feasible to carry out this type of conversion is two-fold: on the one hand one has the advantage of economising on services and energy, as mentioned previously, but in addition, in an SNG refinery the desulphurisation of the heavier products by hydrocracking is fraught with fewer problems than in an ordinary oil refinery. The reason here is that hydrofining to a low sulphur content invariably involves a certain amount of cracking; this might normally limit the intensity of desulphurisation in ordinary petroleum processing but does not impose a similar limit in the case of an energy refinery. In this type of plant, any heavy feedstock that, instead of being merely desulphurised, is also hydrocracked to lighter products becomes perfectly acceptable. After all, the production of SNG and of light fractions that can be easily converted into SNG is the main aim of this operation.

There are again two distinct approaches to the design of an energy refinery. On the one hand, one can hydrofine, hydrocrack and eventually hydrogasify all the products that emerge from the primary fractionation of the crude oil. This results in a configuration shown in the top diagram of Fig. 8.3, where crude oil is fractionated into naphtha and lighter products, light and heavy gas oil and residual fuel oil. The naphtha is hydrodesulphurised and converted into SNG by means of the low temperature reforming process. The light gas oil is hydrocracked and the light products are mixed with the straight-run naphtha reformer feed. The heavy gasoil and residual products, however, are desulphurised in separate facilities, blended and sold as a low sulphur fuel oil.[8,9] Hydrogen, which is required in substantial quantities for this conversion, can be produced either by steam reforming part of the naphtha stream or by partial oxidation of the residual fuel.

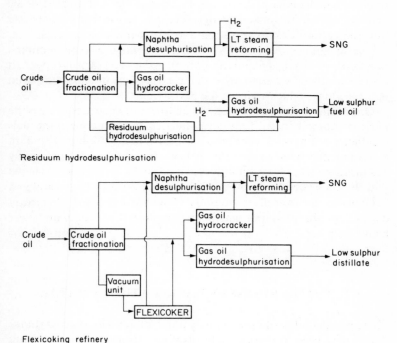

Residuum hydrodesulphurisation

Flexicoking refinery

FIG. 8.3 'Energy' refinery considerations.

The alternative to such a hydrofining–hydrocracking scheme is an energy refinery based on the previously mentioned FLEXICOKING process. A diagram of the refinery operation on this basis is shown in the lower half of Fig. 8.3 and it will be noted that this type of energy refinery consists again of a crude fractionation unit that produces three streams. The light overhead is again hydrodesulphurised and converted into SNG; the middle distillate is again either hydrocracked and then supplies more light feedstock to the SNG unit, or it is merely hydrodesulphurised to produce a low sulphur fuel oil by-product. The difference between this and the previous scheme arises in the treatment of the residual. This is usually vacuum distilled and the distillate is blended with the gas oil side stream from the primary distillation (Fig. 8.1). The vacuum residual goes to a FLEXICOKER where it is converted into light fractions resembling naphtha which undergo the same treatment as the light fraction of the primary distillation and a hydrogen-containing gas that finds its way into the various desulphurisation and hydrocracking systems.

The advantage of the FLEXICOKING configuration is again the fact that no oxygen is required to produce the necessary hydrogen; it also avoids the other alternative, namely hydrogen manufacture from light products, which reduces the SNG yield.

One of the obvious advantages of all forms of energy refining is the fact that all products are fully desulphurised and that none of the fuels produced contains substantial concentrations of sulphur. From an environmental point of view the energy refinery thus becomes highly desirable. As far as its economics are concerned, it can also be shown that these are superior to most other routes to substitute natural gas.

However, the overall profitability of such conversions will clearly depend on the premium which low sulphur fuels command over higher sulphur fuels.

8.4   OTHER REFINERY GASIFICATION SYSTEMS

There are a number of other, mostly older, oil refinery process routes that can be used to prepare light distillates from heavier petroleum products such as distillates and residual fuels. They include catalytic reforming, various forms of thermal cracking, catalytic cracking, visbreaking and deasphalting.[11] With the exception of the last named process they tend to produce olefinic or aromatic materials, which are less suitable for gasification than paraffins.

Furthermore, most of these process steps are designed to increase the yield of motor fuel components, and their economics depend largely on the production of the latter. Little or no attention need, therefore, be paid to these alternatives in the context of SNG manufacture.

8.5 CONCLUSIONS

In the light of the high price of petroleum fractions, and particularly the short supply situation in regard to naphtha and other light products, it is often considered preferable to base future manufacture of SNG on crude oil, fuel oil and other heavy feedstocks. While basically more complex and, of course, also more expensive than reforming or hydrogasification of lighter distillates, heavy oil processes are nevertheless preferable on the grounds of feedstock availability and, if carried out in the form of the so-called energy refinery scheme, also in regard to cost.

A substantial number of crude oil-based SNG plants and energy refinery configurations have, therefore, been proposed, although at the time of writing none of these has actually been built. However, the schemes reported here cannot be considered merely academic exercises in the manufacture of low sulphur fuel oil and substitute natural gas. They have all been costed and complete refinery designs with process integration and design optimisation have been worked out. The results of these studies and the actual costs of manufacturing SNG from heavy feedstocks will be discussed further in Chapter 11 and compared with the economics of other SNG processes.

REFERENCES

1. Hazelton, J. M. and Tennyson, R. N. (1973). SNG refinery configurations, *Chem. Eng. Prog.*, **69**, 97–101.
2. Huebler, J., Janka, J., Seay, G. and Tarman, P. (1973). *Pipeline Gas from Crude Oil, Combined Hydrocracking–Hydrogasification*, Inst. Gas Tech., SNG Symposium I.
3. Knight, W. N. N. and Penniston-Bird, M. L. (1973). Cracking and reforming, in *Modern Petroleum Technology*, 4th Ed. (G. D. Hobson and W. Pohl, eds), Applied Science Publishers, Barking, Chapter 9.
4. Krueding, A. P. (1972). *RCD Isomax Production Route to Today's and Tomorrow's Low Sulphur Residual Fuels*, Am. Inst. Chem. Engs., 71 Nat-Meeting.
5. Lom, W. L. and Agius, P. J. (1975). *Technology and Economics of Clean Fuel Gas Manufacture from Liquid Petroleum*, 9th World Petroleum Congress, Tokyo, P.D. 17, Paper 1.
6. Matula, J. M., Weinberg, H. N. and Weissman, W. (1972). Sour crudes target for new coking process, *Oil Gas J.*, **70**(38), 67.
7. McMahon, J. F. (1973). *The FBH Process—SNG Production from Crude Oil*, Inst. Gas Tech., SNG Symposium I.

8. McWilliams, F. G. and Schuller, R. P. (1973). SNG, naphtha and low sulphur fuel oils from crude, *Proc. Tech. Inst.*, **18**(6/7), 265–266.
9. Morikawa, K., Nojima, S. and Okagami, A. (1975). *Technology and Economics of SNG Manufacture*, 9th World Petroleum Congress, Tokyo, P.D. 17, Paper 2.
10. Janka, J. and Tarman, P. (1974). *SNG from Heavy Oil*, Inst. Gas Tech., SNG Symposium II.
11. Hobson, G. D. and Pohl, W. (Eds) (1973). *Modern Petroleum Technology*, 4th Ed., Applied Science Publishers, Barking.
12. Inst. Gas Tech., Chicago (1972). *Production of Pipeline Gas from Crude Oil Feedstocks*, Report for AGA Inc., Project 6082.
13. Johnson, A. R. *et al.* (1973). *Production of SNG from Crude Oil*, Paper to AIChE Meeting, Louisiana.

Chapter 9

The Conversion of Solid Fuels into SNG

9.1 INTRODUCTION

Technologies for the gasification of coal have been in existence for at least 150 years. However, in spite of this long period of development, there does not exist, to our knowledge, a single full-scale commercial plant in the world that would be capable of producing an entirely interchangeable SNG from a solid feedstock. This apparent paradox is due to a number of historical circumstances, to the complex and varying nature of coal as a feedstock and to the relatively recent development of massive demand for coal-derived fuel gases, *i.e.* for power station and industrial burning purposes.

It is not intended to discuss in great detail the many processes, most of which are now altogether obsolete, that were developed, especially between the two world wars, for the gasification of coal or coke; the aim of most of these was to produce a synthesis gas for ammonia or methanol manufacture or to manufacture a town gas of intermediate calorific value for local distribution to households or minor industries. It is worth mentioning, however, that most of these gasification processes had in common certain technological features such as a low or atmospheric operating pressure, a tendency to produce volatile liquid or sometimes even solid by-products, and to result in a gas that contained substantial amounts of impurities such as sulphur compounds, nitrogen oxides, unsaturated hydrocarbons, referred to sometimes as illuminants, and others. A further feature of these early coal gasifiers was their highly complex, and occasionally inefficient, coal, coke and ash handling equipment.

Owing to the complex nature of coal, it was found necessary, in most early coal gasifiers, to separate the coal carbonisation and char gasification steps. In other words, coals containing a substantial proportion of volatile components were first heated in the absence of air to about 800 °C, so that all volatile hydrocarbons in the coal were expelled and a large proportion of this material was also cracked to lower hydrocarbons and char. This carbonisation of the coal would concentrate between 10 and 40 % of the heating value of the coal in the liquid and gaseous products, leaving between 60 and 90 % in the form of coke or char. Depending on coal structure and quality, this residual carbonised material would be hard and suitable for metallurgical purposes, or soft and friable, and therefore of limited commercial use.

The gasification of this residual char, sometimes in order to balance gas and coke production, became one of the tasks of inter-war technology and led to a number of gasifier designs that operated in conjunction with carbonisation plants although physically separated from the latter. In other instances, and more recently, gasifiers were designed to deal with the entire coal in a single step rather than in two divided gasification processes, an obvious advantage of such processes being the ability to use coal irrespective of the quality of the intermediate coke and of its marketability.

A basic problem of all coal gasification processes, it will be evident by now, is the specific nature of the coal used. Coals differ enormously in physical and, to some extent, in chemical properties and quite often a process designed for one particular grade may have to be modified extensively, or may even be quite unsuitable, for other grades.

Another problem in the gasification of solid fuels is due to the inherent structure of coal. Being a naturally occurring solid, its properties not only vary between different grades, but are also non-uniform within the same deposit. Furthermore, owing to surface effects such as oxidation, moisture, other forms of weathering, etc., there will be large variations in properties between material situated at or near the surface and that taken from the centre of a lump or particle.

One obvious method of overcoming these effects of variability of quality and changes of properties depending on the depth of sampling, is to carry out processing at an extremely slow rate, ensuring that there is complete diffusion of gases throughout the mass of coal, and also uniform heating of particles right to the centre of each individual unit.

Another approach to ensure uniformity of treatment is to prepare the coal by such steps as blending of different grades, pulverisation and sometimes briquetting; all these tend to improve uniformity of penetration, reactivity and behaviour during carbonisation such as swelling or liquefaction.

In fact, one of the main disadvantages of the old coal gasification processes such as carbonisation in horizontal or vertical retorts or coke ovens, water-gas generators, various types of gas producers, etc., is the fact that they used coal with little or no pretreatment. One of the consequences of this approach was that these plants suffered from very low gas production rates per unit of production. What one might call the second generation of coal gasifiers, among which we would include the Winkler, Koppers-Totzek, Rummel and similar generators, used prepared coal and thereby ensured a higher production per unit of output by improving reactivity, for instance by using oxygen instead of air, and also improved penetration rates by the use of fluidised beds, liquid slag beds, and similar devices.

Generalising further, one might say that these attempts to intensify the reaction between gasifying material and coal were further enhanced in a third generation of gasifiers, namely those now under study for the US market, in which output per unit is being further increased by the application of pressure. Examples include the HYGAS, BI-GAS, the CO_2 Acceptor, the molten salt process, etc., all of which are discussed in later sections of this chapter.

Up-to-date gasification processes for the conversion of coal into SNG on a scale aimed at making them competitive with gas from other sources will thus have all or at least most of the following features:[2,10,14,16]

—the coal feed is dried, ground and sized or pulverised before processing,

—its reactivity is enhanced by some form of pretreatment,

—oxygen is used in preference to air to increase reaction rates and reduce gas dilution by nitrogen,

—solids are fluidised by means of gases to facilitate handling, mixing and transfer,

—the main gasification reaction takes place in a fluidised bed or in a high temperature solvent (molten salt or slag),

—the system is pressurised in order to both limit its physical size and produce the gas at pipeline pressure.

It will be noted that certain routes to SNG, particularly the Lurgi process,[5,12] have a number though not all of the features listed above, and although the construction of a substantial number of Lurgi plants has recently been proposed in the US, it is certainly not a third generation gasification process. It is therefore discussed in a separate section in which are also mentioned a number of other fairly advanced coal gasification routes that are fully developed technologically and could be used to manufacture SNG. The main emphasis of this chapter, however, is on the new gasification routes that are being developed in the US with the specific aim of utilising US coal reserves to mitigate the expected shortage of natural gas.

9.2   ESTABLISHED GASIFICATION PROCESSES

The majority of existing processes using coal or lignite as a feedstock and producing liquid fuels, synthesis gas, medium calorific value town gas and lately SNG were developed in Germany just before and during World War II in order to gain independence from imported petroleum fuels. Not all the process routes have been found suitable for SNG production, but schemes based on the Lurgi and Koppers-Totzek routes appear promising.[6]

9.2.1 The Lurgi* Coal Gasification Process
The Lurgi process[12] which is based on the steam–oxygen gasification under pressure of sized coal particles in a mechanically stirred fuel bed is illustrated diagrammatically in Fig. 9.1.

The main reactor is a water-jacketed pressure vessel which also acts as a boiler and supplies the entire process requirement of superheated steam. The coal in the bed is kept in motion by means of a rotating grate that allows solid ash to fall through into a pressurised ash hopper. In addition, stirrer arms reaching into the upper layers of the bed also assist mixing. Fresh coal is introduced batchwise through a similarly pressurised fuel hopper located above the gas generator. Both fuel and ash hopper are equipped with pressure-tight inlet and outlet valves so that either or both can be operated at process or at atmospheric pressure.

* The Lurgi Coal Gasification Process has recently been reviewed from an environmental pollution standpoint. Quantities of solid, liquid and gaseous effluents have been estimated as well as the thermal efficiency of the process. Consequential process modifications have been proposed.[32] Similar studies of the other processes described later in this chapter, viz. Koppers-Totzek, Synthane, CO_2 Acceptor and BI-GAS have also been made.[30,31,33,34]

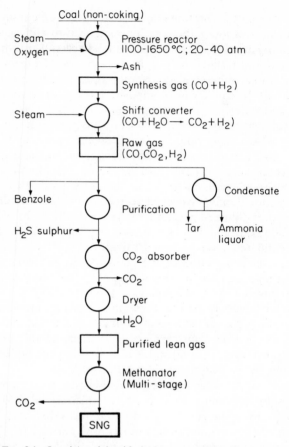

FIG. 9.1   Lurgi (partial oxidation) route to SNG from coal.

While the introduction of coal into the reactor and the withdrawal of ash from the reactor thus present certain engineering problems, the gasifying medium, a mixture of superheated steam and oxygen, is simply injected at a pressure of 20–40 atm into the vessel below the rotating grate, which is thereby cooled, and distributed throughout the coal bed.

The relative proportions of oxygen and steam and the ratio of gasifying medium to coal are strictly defined by a number of factors: excess oxygen will raise the fuel-bed temperature and cause slagging, will tend to melt the rotating grate, and is undesirable economically; excess steam will lower the reaction temperature and decrease gas production; insufficient steam/oxygen will result in incomplete gasification and residual carbon in the ash; use of excess gasifying medium is uneconomic, raises reactor operating temperature and tends to shift the equilibrium composition of the gas towards carbon dioxide and water.

The reactions taking place in the Lurgi gas generator are the usual ones of coal carbonisation, *i.e.* expulsion of volatile hydrocarbons from the coal and their subsequent cracking to methane and lower hydrocarbons; the synthesis gas reaction of the coke or semi-coke formed by carbonisation with steam and oxygen, with the result that carbon monoxide and hydrogen formed from the coal contribute largely to the volume of gas; and finally the methanation reaction of carbon monoxide and hydrogen under pressure. The gases, although formed at different levels in the reactor, are collected jointly and piped to the purification section of the plant. Before entering the latter, however, they are cooled in a steam generator which contributes to the overall steam output of the plant.

The cooled gas is passed through a carbon monoxide shift convertor where part of the carbon monoxide reacts with excess steam and produces hydrogen and carbon dioxide. Tar and ammonia liquor are withdrawn from the condensate, both in the steam boiler and in a cooler after the carbon monoxide shift.

The next stage of gas purification consists in the removal of aromatics by scrubbing with 'benzole' which is allowed to flow countercurrent to the gas down a column, recovered by distillation and recycled. The aromatics-free gas is now purified from organic sulphur compounds and hydrogen sulphide by passing through an alkazid absorption plant. Sulphur can be recovered from the scrubbing liquid and the lean alkazid fluid is returned to the plant. The next stage of purification consists in the removal of residual sulphur compounds in iron oxide boxes and is followed by carbon dioxide removal in a CO_2 absorber. Various types of equipment, *i.e.* Benfield, Vetrocoke or Catacarb, can be used for this service. Final purification of this lean gas consists in water removal and drying in a glycol absorption tower.

Table 9.1 lists the composition and properties of gas produced from Scottish bituminous coal at the various stages of the Lurgi process. It also shows that by the addition of 5·8 mol % butane (and some air)

TABLE 9.1
Typical Lurgi Gasifier Gas Characteristics

	Crude gas (H_2S-free basis)	Purified gas (CO converted)	Butane added (UK town gas quality)	Methanated – SNG
Composition:				
mol % CO_2	25·6	2·1	1·5	1·0
O_2	—	—	0·2	—
C_4H_{10}	—	—	5·8	—
C_nH_m	0·6	0·4	0·6	0·6
CO	24·4	9·8	6·9	0·1
CH_4	10·3	13·6	9·6	95·1
H_2	37·3	71·8	50·4	0·7
N_2	1·8	2·3	25·0	2·5
Combustion characteristics				
Calorific value (gross), Btu/scf	312	404	477	953
Relative density, 15°C, 1 atm, (air = 1)	0·738	0·280	0·549	0·558
Wobbe Index	—	—	642	1 289
Weaver flame speed factor	—	—	34·4	14·9

this gas can be converted into a standard town gas meeting the rather strict UK gas specifications.

Clearly, conversion of the gas described in the second column into SNG will require more than simple addition of butane to enrich the gas to the necessary calorific value. The approach that is likely to be adopted in all the major Lurgi plants being prepared for the US market is to methanate this gas to about a 95 % vol methane content. (The details of methanation processes used for coal gas enrichment will be discussed in the following chapter.) The last column in Table 9.1 lists the composition of such a gas after multi-stage methanation and CO_2 reduction to a content of 1 mol %.

While Lurgi gas methanation can be shown in theory to be feasible and the equipment for this purpose can be designed with a reasonable degree of confidence, it was felt that an experiment in methanation of this gas on an industrial scale would be desirable and a demonstration plant was built alongside the Westfield Lurgi plant near Edinburgh.[8] The results of the work carried out at Westfield have confirmed that Lurgi gas can effectively be converted into SNG and that the design of large commercial plants for the United States would appear to be satisfactory.*

Gasification conditions in Lurgi generators are of the order of 1 tonne of coal per hour per m^2 of grate area—they range from 500 to 1500 kg depending on type of coal and also to some extent on plant pressure. At 80 % gasification efficiency this corresponds to an output of about 10 million scf/day of SNG per 3 m diameter Lurgi gas generator. In other words a typical US plant producing 125 million scf of SNG per day will consist of a minimum of 13, in actual fact probably 15 or 16, generators. The possibility of building units larger than 3 m interior diameter, of course, exists. However, there is some doubt whether efficient rotating grates and stirrer mechanisms of such dimensions can be designed. It also follows that such generators could no longer be fabricated in a shop and shipped to the site but that far more on-site manufacture would have to be introduced. Furthermore, channelling and irregular gas flow, while already a risk in present designs, would be even more likely for large diameters.

One of the main drawbacks of Lurgi gasifiers is thus their limited throughput per unit of plant and the consequent necessity to operate multiple generators in parallel. This carries the penalties of a higher

* At the present time, SNG produced at Westfield is mixed with natural gas in a 3600 Mcfd distribution grid with SNG representing about 60 % of the gas consumed.[29]

operating labour requirement and of more frequent mechanical breakdown. The latter is an important consideration in plant of the complexity of a Lurgi generator, which has far more mechanically exposed moving parts and also more high-pressure valves controlling the flow of solids, refractory linings exposed to high temperature, moving solids and other vulnerable features, than has the average chemical process plant.

In addition, Lurgi gasifiers are, of course, surrounded by the usual paraphernalia of large coal processing industries such as coal storage yards, cranes and gantries, conveyors, lifts and unloading facilities for coal and ash disposal. The gas cleaning section of the plant is similarly complex and consists, apart from the scrubbers and wash columns already mentioned, of solvent recovery equipment and the processing of by-products such as benzene and other aromatics, hydrogen sulphide and sulphur, carbon dioxide and condensate. It will be appreciated, however, that the need to deal with coal gasification by-products is, on the whole, not a function of the gasification process, and practically all coal gasifiers will require equipment of the same type and size to produce a similar volume of SNG.

9.2.2 Other Commercial Coal Gasification Systems

While quite different technically, a number of other coal gasifiers were developed at about the same time as the Lurgi process and are now commercially available for immediate use. They include the Winkler generator which was one of the first commercial applications of the fluid-bed gasification technique.[1] Small coal or coke particles (average diameter 0·8 mm) are gasified at atmospheric pressure in a steam/oxygen blast, the fuel ash being carried out of the reaction zone by the gas stream. The process tends to be inefficient owing to incomplete separation and the carry-over of fuel rather than ash. Also, operating at atmospheric pressure, it has a limited output per unit of production.

The principle of the Koppers-Totzek[6] process is similar. However, instead of carefully avoiding slag formation by controlling the reaction temperature of the fluid bed and using coals of high melting point ash, this process is designed to produce liquid slag by steam/oxygen combustion of a number of jets of finely divided carbon in a refractory chamber. Liquid slag drains away while hot gases pass through a vertical waste heat boiler, to ensure return to the combustion chamber of entrained fuel and some of the slag. The higher temperature, compared with the Winkler generator, ensures complete cracking of tar and other liquid products and thus simplifies gas treatment and purification. A high pressure version, along the lines of the Shell fuel oil gasifier is said to be under development.

Other gasifiers that rely entirely on liquid slag disposal and operate at correspondingly higher temperatures are the Rummel single and double shaft designs in which steam/oxygen, or steam and air, are injected tangentially into a bath of molten slag and act both as gasifying medium and as a source of mixing energy. In the double shaft the air blast heats the bath while steam injection produces fuel gas.

An obvious disadvantage of slagging and other high temperature gasifiers as far as SNG manufacture is concerned is the very low content, if not complete absence, of equilibrium methane and the consequent need for additional methanation capacity.

9.3   NEW CONCEPTS OF COAL GASIFICATION

The declining natural gas reserves in the United States, in conjunction with the availability of very large coal reserves, particularly in areas relatively remote from existing industrial development, have provided the incentive for the proposed construction of very large, on average 250 million scfd, gasification plants close to new coal mines that are to be opened up in the states of New Mexico and Arizona. While technology to convert these coals into pipeline gas is available, it has been shown in the previous section that improvements in technique and especially in the economics of coal gasification would appear desirable. Very intensive development work on new gasification processes, which would overcome some or all of the drawbacks of the Lurgi process and some of the other older conversion routes, is now under way; a number of pilot plants have already been built and the first of the 'second generation'* coal gasifiers should be on stream in the early 1980s.

While it is difficult to forecast with certainty which designs amongst those now under development will in fact be used commercially, the

---

* In this context, the Lurgi–Koppers-Totzek and Rummel gasifiers are regarded as representing a first generation. Yet a 'third generation' including Exxon Research's Catalytic Gasification Process is also under development as a more distant prospect.

following discussion deals with processes that are considered promising and that would appear to meet the technical and economic requirements of the near future.

9.3.1 The HYGAS Process[15]

This process is being developed by the Illinois Institute of Gas Technology in Chicago and is based on the hydrogenation of a fluidised bed of specially prepared coal. The latter is introduced as a slurry into the top layer of three superposed fluidised beds. Hot hydrogen enters the bottom bed and fluidises all three beds, evaporating at the same time the oil from the slurry that was injected into the top. The fluid bed gasifier is maintained at a pressure of 75–90 atm and this is conducive to the formation of methane during the coal gasification process.

The ungasified portion of the coal is obtained in the form of a reasonably reactive type of char which is withdrawn from the bottom bed and transferred into a hydrogen generator. The latter is normally operated by injecting a steam/oxygen mixture into the fluid char, but can, in the absence of oxygen, operate on steam alone provided the bed is electrically heated. In fact the plant now operating at the IGT research centre is using this version of the process. A third possible method of hydrogen production, based on the steam iron reaction and a subsequent reduction of the iron by means of char, is also being considered.

The hydrogen gas is produced under process pressure and therefore does not require recompression when it enters the lowest of the three coal beds. The reason for arranging the gasification reaction in three stages is the possibility of obtaining a more favourable approach to methane equilibrium in this fashion. Temperatures of the gasifier range from 1050 °C in the hydrogen generation stage to 316 °C in the coal drying and preparation zone. Pressures are uniformly the same in the entire equipment and depend on the gas pipeline pressure required.

The late 1974 status of the process is that of a demonstration plant processing 75 t a day already in operation with some of the auxiliary plants, such as hydrogen generator and coal preparation, still under construction. A flow diagram of the HYGAS process is shown in Fig. 9.2.

SNG produced in the HYGAS unit is claimed to be fully

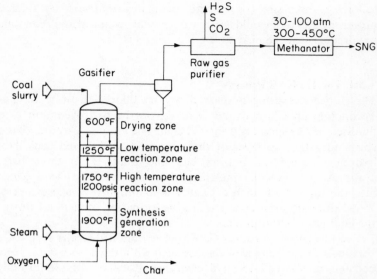

FIG. 9.2   HYGAS process. (Source: Ref. 2.) (Courtesy: Institute of Gas Technology, Chicago.)

interchangeable with US natural gases in US appliances provided excess hydrogen has been removed by one or two methanation steps.*

9.3.2 The BI-GAS Process[21]

This fluid-bed process is being developed by Bituminous Coal Research Inc. and at the time of writing had reached a stage when a large (120 ton/day) pilot plant was nearing completion.

The BI-GAS system consists of two stages of gasification operating at about 930 °C and 1670 °C respectively. Coal is introduced as a coal–water slurry into a heat exchanger/separator where it is dried and preheated by the raw gas made in the process. The dried coal powder, together with steam, enters the low temperature section of the gasifier where it is devolatilised and partly gasified by a stream of hot reducing gas.

Char formed in the process is carried overhead into a cyclone

* Early 1975, it was announced that a successful continuous run had been conducted with the inclusion of the steam/oxygen plant for converting the residual char into hydrogen. The product gas (N_2-free basis) contained 96·5 % vol methane, 3·5 % vol hydrogen and had a heating value of over 970 Btu/ft^3.[28]

separator from which it is returned to the high temperature gasifier. The latter is also charged with steam and oxygen; the resultant high temperature melts all ash components and coagulated slag runs down the walls of the reactor; it is allowed to drop into a pool of quench water and is periodically withdrawn as a solid powder or sludge.

The process is characterised by the wide range of coals that can be gasified irrespective of their volatiles content and mineral ash. It can be operated over a fairly wide range of pressures, which can be as high as 75 to 90 atm, and is thought to be one of the prime contenders for early commercialisation. A flow diagram of a BI-GAS coal gasifier is shown in Fig. 9.3.

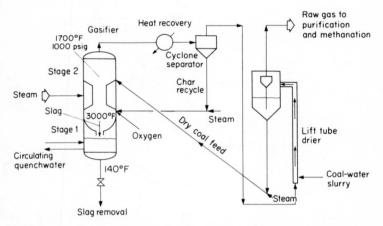

FIG. 9.3   BI-GAS process. (Source: Ref. 2.) (Courtesy: Institute of Gas Technology, Chicago.)

Raw gas produced in a BI-GAS plant is not interchangeable with US natural gases; however, it can be converted, usually by two successive methanation steps, into a fully interchangeable product.

9.3.3 The CO₂ ACCEPTOR Process[7]

This approach to coal gasification is being developed by the Consolidation Coal Co. A 40 ton per day pilot plant has been in operation since 1973 and a number of different types of lignite and sub-bituminous coal have been tested.

The process is best explained by reference to the sequence of intermediate steps given in Fig. 9.4a. It differs from other routes to

SNG in that the endothermic heat of reaction required to gasify coal with steam is provided by simultaneously reacting calcined dolomite, a mixture of calcium and magnesium oxides, with carbon dioxide. By doing this the need for oxygen or external heat is obviated, and pulverised lignite fluidised by means of steam and recycle gas can be gasified in two separate stages of fluid-bed processing, *i.e.* in the 'devolatiliser' and the 'gasifier' section of the plant respectively. Both fluid beds are heated by means of the above mentioned chemical

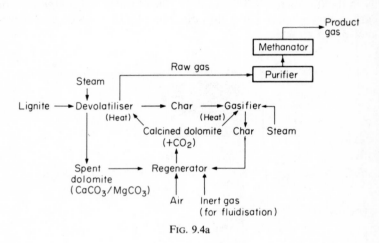

FIG. 9.4a

reaction. The dolomite is regenerated by calcination in a third vessel (the 'regenerator'), again in a fluidised bed, which is heated by firing with completely devolatilised char and air, both injected under medium pressure (20 atm).

Calcined dolomite overflows from the regenerator into the devolatiliser and gasifier beds respectively, and reacts in each with carbon dioxide in the gas; spent 'acceptor', *i.e.* reformed dolomite, is withdrawn from the bottom of the two beds and raised by means of inert gas into the regenerator. Char is collected internally from the top of the 'devolatiliser' fluid bed and transferred into the gasifier, where it reacts with steam, again assisted by the exothermic heat of re-combination. Residual char in the 'gasifier' is similarly withdrawn from the surface of the fluid bed and introduced into the regenerator where it is burnt, as previously explained.

Gas taken overhead from the gasifier is recycled for char

fluidisation processes to the devolatiliser, from whence raw gas is withdrawn, purified and methanated.

Coal ash is eliminated by entrainment overhead with the regenerator combustion gases, although no doubt some ash adheres to the cycling dolomite and small quantities will be found in the raw gas before purification and methanation. Combustion air and inert gas for the transfer of solids are supplied under pressure. Steam is generated by heat exchange with hot raw gas and regenerator exit gas.

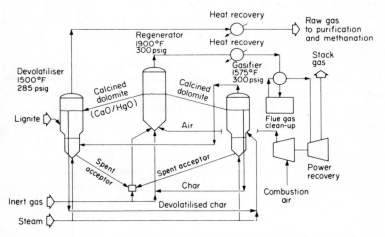

FIG. 9.4b CO_2 acceptor process. (Source: Ref. 2.) (Courtesy: Institute of Gas Technology, Chicago.)

Pressures in the Consolidation CO_2 Acceptor system are lower than those in the HYGAS and other gasification processes—20 atm in regenerator and gasifier, slightly less in the devolatiliser—but high enough to limit the size of the equipment. It is not easy to pressurise to a higher pressure because of its air requirement. Temperatures are 815 °C in the devolatiliser bed, 860 °C in the gasifier and 1040 °C in the regenerator.

A flow diagram of the CO_2 Acceptor process is shown in Fig. 9.4b.

It is difficult to produce a fully interchangeable SNG by this route since improvement potential by methanation has practically been exhausted. The gas produced so far therefore tends to fall, albeit close to the boundary, just outside the recognised interchangeability area.

9.3.4 The Synthane Process[3]

This route to SNG is being worked on by the US Bureau of Mines, and a 70 ton/day pilot plant went on stream at the end of 1974. The process was designed for the conversion of a large variety of coals, and to ensure that coking coals do not result in bogging and other forms of maloperation of the fluid bed, such coals are pretreated before injection.

This is done in similar fashion as in the BI-GAS process by slurrying the pulverised coal with water, drying it in a lift tube drier by contact with the gasifier overhead and separating dry coal and raw gas in a cyclone vessel. Raw coal then enters the gasifier through a stand pipe where it is reacted at a temperature of about 980 °C with steam and oxygen in a stationary fluid bed.

An internal cyclone returns coal and other solids entrained with the gas back into the bed. Char is continuously withdrawn from a conical bottom extension of the bed and quenched with boiler feed water, producing steam required for gasification. This is subsequently superheated by heat-exchange with the overhead gas. The latter is further cooled in the coal drier and eventually quenched before entering the purification and methanation section.

The Synthane process is designed to operate at pressures of 40–75 atm. It is claimed to have high gasification rates due to coal pretreatment and careful design of the fluid bed, which avoids local hot spots in spite of the high gasification temperature.

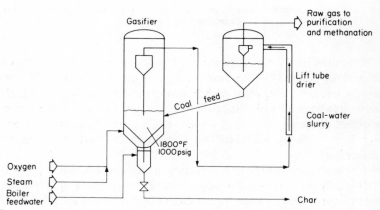

FIG. 9.5   Bureau of Mines Synthane Process. (Source: Ref. 2.) (Courtesy: Institute of Gas Technology, Chicago.)

On the other hand, this route may present problems in regard to the extent of gasification: the residual char withdrawn from the fluid bed is not treated any further, and although the quantities involved are said to be small, provision must be made for char handling and disposal, generally by firing in a steam boiler or power plant.

A flow diagram of the Bureau of Mines Synthane process is shown in Fig. 9.5.

9.3.5 The Agglomerated Ash Process

This route is being developed in co-operation by Union Carbide and the Battelle Research Institute. It is another high temperature fluid-bed process but avoids the use of oxygen by the device of a separate char burner in which char and ash are oxidised by means of compressed air. The development of the process has reached a stage where a 25 ton/day pilot plant was due to start operations towards the end of 1974. The route is suitable for most bituminous coals since it incorporates a steam/air pretreatment stage for the reduction of the coking tendency of certain coals.

The name, agglomerated ash process, is due to the method used to correct the heat deficiency of the fluidised-bed gasification reaction. Char is withdrawn from the surface of the high temperature (980 °C) fluid bed while agglomerated ash, formed in the unusually deep bed of the gasifier, is drained from the conical bottom section of the reactor. The mixture of char and ash is introduced, together with combustion air, into the burner vessel and hot (almost 1100 °C) agglomerated ash particles are transferred from the surface of the burner fluid bed into the gasifier.

No air or oxygen enters the gasifier and both fluidisation of the bed and gasification are effected by means of steam, the latter being generated by heat exchange with raw gas leaving the gasifier, and with hot flue gas at the exit of the burner. Ash in excess of that required for heat transfer is withdrawn from the cold (815 °C) agglomerated ash line.

The process, in line with other routes requiring compressed air, is carried out at relatively low pressure (about 7 atm), and since its gasification temperature is high, little or no methane is formed. Methanation duty of the gas after-treatment facilities, if SNG is to be the product, is consequently high. However, there are practically no liquid or unsaturated by-products and there is no need for tar, aromatics or carbon black separation.

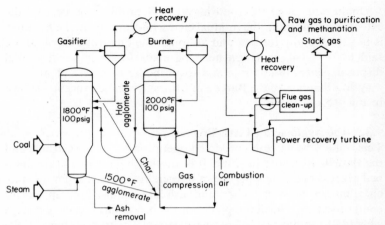

FIG. 9.6   Battelle–Union Carbide Agglomerated Ash Process. (Source: Ref. 2.)
(Courtesy: Battelle Research Institute.)

A flow diagram of the Battelle–Union Carbide agglomerated ash coal gasification process is shown in Fig. 9.6.

9.4   OTHER COAL GASIFICATION DEVELOPMENTS

In addition to the Lurgi process, which is commercial, and the five US processes that are being tested in large pilot plants (HYGAS, BI-GAS, CO_2 Acceptor, Synthane and Agglomerated Ash Process) there are a number of processes being developed in the US which may be transferred from the bench or laboratory scale to pilot plant testing in the near future. They include:

The molten salt process.[4] This process is under study by M. W. Kellogg and uses a medium of molten sodium carbonate at 950 °C in the gasification of pulverised coal at a pressure of 75–80 atm by means of steam and oxygen. The catalytic effect of the sodium carbonate is claimed to result in the breakdown of higher hydrocarbons, while the methane-char equilibrium remains unaffected. Ash is removed by purging some of the melt.

The Hydrane process.[24] This is another US Bureau of Mines development, in which coal is reacted with hydrogen during free fall, at a temperature of 900 °C and a pressure of between 35 and 70 atm. The resultant char is gasified with steam and oxygen in a fluid bed to produce the required hydrogen. It is claimed that all types of coal can be gasified without pretreatment.

The molten iron route.[11] This is advocated by Applied Technology Corp. A molten iron bath is used as the gasification medium and complete gasification of most types of coals is achieved by blowing steam and oxygen through the bath. Ash and sulphur compounds are removed by lime addition and skimming of fluxed slag from the bath.

Advanced fluid-bed gasifier.[19] This process is being developed by the Westinghouse Co. in the USA. As in the CO_2 Acceptor process, coal is devolatilised and then gasified in two separate stages. In the first stage, dolomite is added with the crushed coal to remove sulphur from the hot gases produced. Fluidisation of the coal in the devolatiliser is brought about by means of gas recycled from the second or gasification stage where devolatilised char is gasified using steam and oxygen at $100\,°C$, 10–30 atm pressure. The product gas is taken off from the top of the devolatiliser.

9.5   CONCLUSIONS

Modern coal gasification processes tend to combine the de-volatilisation of coal with at least partial gasification of the resultant char. Both devolatilisation and char gasification by means of steam are endothermic and their heat requirements must be met by burning part of the coal. This is normally done with the aid of oxygen; however, the cost of air separation is high and a number of attempts have been made either to use air and withdraw the combustion gases separately to avoid dilution of the product gas, as in the Agglomerated Ash or Rummel two-shaft slagging generator, or alternatively to utilise other sources of heat energy. In the CO_2 Acceptor process, for example, one utilises the heat of reaction between carbon dioxide and calcined dolomite; in one version of the HYGAS process the reaction between char and steam is advanced by electrical heating.

A different approach towards balancing the heat requirement of coal carbonisation and gasification is the use of hot high pressure hydrogen to initiate the methanation of char, which is exothermic, simultaneously with the devolatilisation of the coal. This technique is used in the HYGAS and Hydrane routes and has the advantage of producing a raw gas containing some methane and, therefore, of requiring less intensive methanation to produce SNG. However, the hydrogen must be prepared, generally by reacting some of the char with steam and oxygen and treating the water gas to raise its hydrogen content.

The need to transfer large quantities of hot solids rapidly and effectively has been mentioned, and fluid-bed techniques have been found effective. They also result in rapid mixing, higher gas–solid reaction rates and uniform temperatures throughout the system. Thus stationary beds with solid overflow or bottom withdrawal are used in the HYGAS, CO_2-Acceptor, Synthane, Hydrane and Agglomerated Ash processes. Entrained flow, dilute bed (transfer line reactor) techniques are used in the BI-GAS and in two stages of the Hydrane process.

Two other methods of transferring solids have also been used for coal gasification: the Lurgi generator uses a moving bed, and beds of molten materials are encountered in the Kellogg molten salt system and the Applied Technology molten iron gasifier.

Operating pressures in the gasifiers discussed in this chapter vary widely. A high operating pressure, it has been mentioned, is desirable because it results in higher methane concentrations in the gas and also because it tends to increase the output of gas per unit. Oxygen is normally available under pressure and oxygen gasifiers are therefore usually designed to operate at higher pressures than do air oxidation units, which require a higher air compression energy if a high gas outlet pressure is required.

Operating temperature in a gasifier has an effect on both coal solids and outlet gas composition, apart from the need for special materials of construction if temperatures are unusually high. Coal, depending on type and quality, will melt or cake at certain temperatures and coal ash will coagulate and ultimately produce a liquid slag. Design of a gasifier must be such that gasification proceeds sufficiently fast to prevent caking—many coals require pretreatment for this purpose—and maximum temperatures must be carefully controlled if ash is to be removed in solid form. If there is provision for liquid slag removal, such as in the BI-GAS or molten iron routes, a minimum operating temperature will ensure that ash is always present in liquid form.

The effect of a high outlet temperature on gas composition is a lowering of methane content; however, liquid products such as tar and aromatic oils, which complicate the aftertreatment of the raw gas, are effectively broken down to lower hydrocarbons, char and hydrogen, which are more easily handled than the former.

The introduction of the solid fuel into the gasifier is almost invariably in the form of pulverised coal. However, if the internal pressure is very high—over 70 atm—it becomes difficult to introduce the solid fuel through a pressurised lock hopper such as that of the Lurgi gasifier, and a water–coal slurry becomes the preferred alternative, e.g. in the HYGAS and BI-GAS processes. Coal is processed as a very fine powder in all modern gasification routes with the exception of the Lurgi and the molten medium processes. However, the particle size varies from fairly large in the Hydrane free-fall reactor to very fine in the BI-GAS and Synthane fluid beds.

A comparison of the different gasification routes discussed in this chapter with reference to the various criteria mentioned above will be found in Table 9.2, which summarises published information on gasification reactor, medium, temperature and pressure for the six processes that have reached the pilot plant stage. Table 9.2 also gives some indication of the gas quality produced under those conditions, and of the type of coal that can be gasified.

A final decision as to which routes will ultimately be used to produce SNG commercially has obviously not been made at this stage. The only recent orders, so far, have been for Lurgi plants, although the US utilities who placed them were fully aware of the somewhat obsolete character of the process and its high investment cost, reasoning that it was at least fully proven in its critical process sections. The fact that a series of modern processes are under test is evidence that future coal gasifiers will probably use different design principles, and while selection of the most promising route is mostly a matter of opinion at the present time, our guess is that the CO_2 Acceptor process, because of the absence of oxygen, the elegant methods of handling and transference of solids and its ability to gasify lignites and low grade coals, will ultimately lead the field. However, the field is still open and much additional development work will be required before utility companies will make their final choice. Amongst the hurdles still to be overcome are: the opening up, production and handling of the very large quantities of coal required to produce sizeable quantities of SNG (15 000–25 000 tons of coal per day for 250 million scf of SNG per day[19]); the availability of the requisite volume of water for steam raising purposes; environmental

TABLE 9.2
Characteristics of Coal Gasification Processes

Process	Coal			Gasification				Raw gas characteristics
	Types	Size	Pretreatment	Reactor	Medium	Temperature (°C)	Pressure (atm)	
Lurgi	non-coking	3–25 mm	none	moving bed	steam/O_2	700–1 200	25–40	tar, high CH_4
BI-GAS	all	200 mesh	none	2 stage, entrained flow	steam/O_2	1 480	70–100	little CH_4
CO_2 Acceptor	lignite, sub-bitum.	—	none	2 stage, fluid bed	steam	870	20	little CH_4
HYGAS	non-coking	—	none[a]	3 stage, 2 fluid beds	steam/H_2	900	70–100	some CH_4
Synthane	all[b]	240 mesh	yes	fluid bed	steam/O_2	980–1 040	40–70	some CH_4
Batelle–Union Carbide	all[b]	8–100 mesh	yes	fluid bed	steam	980	7	no CH_4

[a] Needs pretreatment for coking coals.
[b] Provided pretreatment is effective.

protection; reduction in overall plant size and of investment costs; continuing availability of cheap coal.

REFERENCES

1. Banchik, I. N. (1973). *The Winkler Process*, Paper 8, Inst. Gas Tech., Clean Fuels from Coal Symposium.
2. Bolln, John P. (1973). *Commercial Concept Designs*, 5th Synth. Pipeline Gas Symposium, Chicago.
3. Corder, W. C., Batschelder, H. R. and Goldberger, W. M. (1973). *The Union Carbide/Battelle Coal Gasification*, 5th Synth. Pipeline Gas Symposium, Chicago.
4. Cover, A. E., Schreiner, W. C. and Skaperdas, G. T. (1973). Kellogg's coal gasification process, *Chem. Eng. Prog.*, **69**(3), 31–36.
5. Elgin, D. C. and Perks, H. R. (1973). *Trials of American Coals in Lurgi Pressure Gasification Plant at Westfield, Scotland*, 5th Synth. Pipeline Gas Symposium, Chicago.
6. Farnsworth, F. J., Leonard, H. F., Mitsak, D. M. and Wintrell, R. (1973). *The Production of Gas From Coal Through a Commercially Proven Process*, Koppers Co. Inc.
7. Fink, C. E. (1973). *The CO$_2$ Acceptor Process*, Paper 16, Inst. Gas Tech., Clean Fuels from Coal Symposium.
8. Hebden, D. (1973). Production of SNG from coal, *Gas World*, **CLXXVIII**, 341.
9. Hegarty, W. P. and Moody, B. E. (1973). Evaluating the BI-GAS SNG process, *Chem. Eng. Prog.*, **69**(3), 37–42.
10. Huebler, J. (1973). *Coal Gasification, State of the Art*, Paper 22, Inst. Gas Tech., SNG Symposium I.
11. La Rosa, P. (1973). *Fuel Gas from Molten Iron Coal Gasification*, Paper 15, Inst. Gas Tech., Clean Fuels from Coal Symposium.
12. Moe, J. M. (1973). *SNG from Coal via the Lurgi Gasification Process*, Paper 4, Inst. Gas Tech., Clean Fuels from Coal Symposium.
13. Quade, R. N. and McMain, A. T. (1973). *Nuclear Energy for Coal Gasification*, Paper 22, Inst. Gas Tech., Clean Fuels from Coal Symposium.
14. Reboul, J. (1974). Les techniques nouvelles de production de gaz et d'hydrocarbures liquides à partir du charbon, *Rev. Française de l'Energie*, 517–528.
15. Schora, F., Lee, B. S. and Huebler, J. (1973). *The HYGAS Process*, Paper 12, Inst. Gas Tech., Clean Fuels from Coal Symposium.
16. Schreiner, W. C. and Finneran, J. A. (1974). *Trends in the Development of SNG Processes*, Symposium on Future Gas Supplies, Amsterdam.
17. Siegel, H. M. and Kalina, T. (1973). Coal gasification costs may lower, *Oil Gas J.*, **71**(7), 87–94.
18. Wett, T. (1973). SNG from coal involves big projects, *Oil Gas J.*, **71**(26), 131–4.

19. Henry, J. P. Jnr and Louks, B. M. (1975). *Coal gasification in the United States*, 9th World Petroleum Congress, Tokyo, P.D. 17, Paper 5.
20. Klingman, G. E. and Schaaf, R. P. (1972). Make SNG from coal? *Hydrocarb. Process.*, **51**(4), 97.
21. Grace, R. J. *et al.* (1973). *Design of BI-GAS Pilot Plant*, Paper to 5th Synthetic Pipeline Gas Symposium, Chicago.
22. Propert, P. B. (1973). *Design and Fabrication of BI-GAS Reactor*, Paper to 5th Synthetic Pipeline Gas Symposium, Chicago.
23. Lee, B. S. (1973). *Status of HYGAS Process*, Paper to 5th Synthetic Pipeline Gas Symposium, Chicago.
24. Feldmann, H. F. and Yavorsky, P. M. (1973). *The Hydrane Process*, Paper to 5th Synthetic Pipeline Gas Symposium, Chicago.
25. Cairns, J. and Hartill, I. J. (1973). *British Gas Experience of High Pressure Gas-Making Processes*, Paper to 5th Synthetic Pipeline Gas Symposium, Chicago.
26. Anon (1971). Gasification of coal, in *Encyclopaedia of Chemical Technology*, 2nd Ed., Supplement volume, p. 198, Wiley, New York.
27. Paquette, A. J. and Beychok, M. R. (1974). Coal gasification for clean energy, *Energy Pipelines and Systems*, March, 48.
28. Institute of Gas Technology (1975). *Gas Scope*, No. 31, Chicago.
29. Anon (1975). Squeezing oil and gas from coal, *Energy* (Chilton Co.'s Oil and Gas), p. 34.
30. Office of Research and Development, US Environmental Protection Agency. Report GRU, 3DJ.74, Koppers-Totzek Process.
31. Office of Research and Development, US Environmental Protection Agency. Report GRU, 4DJ.74, Synthane Process.
32. Office of Research and Development, US Environmental Protection Agency. Report GRU, 5DJ.74, Lurgi Process.
33. Office of Research and Development, US Environmental Protection Agency. Report GRU, 6DJ.74, CO_2 Acceptor Process.
34. Office of Research and Development, US Environmental Protection Agency. Report GRU, 9DJ.75, BI-GAS Process.

Chapter 10

Industrial Methanation Processes

10.1 INTRODUCTION

The methanation of carbon monoxide and carbon dioxide, the thermodynamics and kinetics of which reactions* have been discussed in Chapter 5, is an important processing step in the conversion of liquid and solid fuels into SNG. Usually one finds that in the preliminary stages of SNG manufacture, apart from methane, a number of lower calorific value gases are produced simultaneously in the various types of gasification reactions that were discussed in earlier chapters. Thus oxygenation and steam reforming both produce carbon oxides, the calorific value of which is about $300 \, \text{Btu/ft}^3$ for carbon monoxide and, of course, zero for carbon dioxide; hydrolysis generally leaves some residual hydrogen in the gas mixture and this also has a heating value of around $300 \, \text{Btu/ft}^3$.

In order to convert these low calorific value gases into methane the carbon oxides have to be reacted with hydrogen; this means either that extra hydrogen must be introduced or that one may find oneself left with surplus carbon oxides, of which only carbon dioxide can be removed, usually by absorption in alkaline absorbents. It is therefore clearly important that at least all the carbon monoxide present must be methanated before further processing of the gas.

Methanation, as an industrial process, has been known for many years since it is an essential gas purification step in the synthesis of ammonia from hydrogen and nitrogen. Residual carbon monoxide, which is a potent poison for all ammonia catalysts, has to be completely removed from the synthesis gas stream, an undertaking that is not too difficult in the light of the large surplus of hydrogen present. Usually carbon dioxide has been completely removed from the gas stream before it reaches the methanation section of the plant.

* $CO + 3H_2 \rightleftharpoons CH_4 + H_2O$; $CO_2 + 4H_2 \rightleftharpoons CH_4 + 2H_2O$.

However, small concentrations of CO_2 do not affect the process and only when centrifugal ammonia syngas compressors were introduced did the presence of heavy gas components become undesirable.

Methanation as a step in ammonia manufacture, or in the production of pure carbon-monoxide-free hydrogen, is thus characterised by the presence of excess hydrogen, by a low concentration of carbon monoxide and the almost total absence of carbon dioxide (*see* Table 10.1), and by a moderate pressure of up to 25 atm. The synthesis gas has usually been cooled and is therefore almost dry.

Methanation of first stage town gas to produce SNG, on the other hand, usually has to deal with much higher concentrations of carbon dioxide and carbon monoxide, generally with the former predominating. The result of methanating such a gas is the presence of residual carbon dioxide, which must subsequently be eliminated or reduced by alkaline scrubbing and thus necessitates a catalyst that preferentially hydrogenates carbon monoxide. The catalyst should also be steam resistant since wet recycle gas is commonly added as a diluent.

Differences between commercial methanation to produce ammonia synthesis gas or hydrogen on the one hand, and SNG on the other, are not the only ones that need be considered here. One must also distinguish between situations such as the relatively low concentration of hydrogen (10–20% vol) in a low temperature reformer exit gas and the much higher hydrogen content of a gas produced by hydrogasification of heavy petroleum fuels. Finally, gases produced by the processing of coal will again differ in composition, and particularly in impurity content, from the gases considered so far, and will also impose their particular restrictions on the methanation process selected for their enrichment to SNG.

The second, third and fourth columns in Table 10.1 list a number of gas compositions from different sources which one may wish to convert into SNG. Each of the gases is basically suitable for enrichment but different catalysts and different operating conditions such as process pressure, inlet temperature, flow rate, dilution, intermediate cooling, etc. apply.

In the light of these differences it is proposed to discuss first methanation catalysts that are suitable for working with the different gas compositions and operating conditions and to review in somewhat greater detail in subsequent sections, variations of the basic

TABLE 10.1
Intermediate Gases to be Methanated

	Ammonia synthesis gas	Gas ex LT reformer	Gas ex GRH/CRG system	Typical coal-based SNG
Composition (mol %)				
Nitrogen	24·4	—	—	—
Hydrogen	74·2	12·1	19·5	53·6
Carbon monoxide	0·6	0·6	0·9	17·8
Carbon dioxide	0·05	11·1	16·8	—
Methane	0·4	30·0	52·8	13·6
Steam	0·3	46·1	10·0	15·0
Sulphur	nil	nil	nil	depends on gasification route

methanation process that have been used or proposed for the various SNG systems under consideration. Among the latter are the low temperature reforming step followed by double methanation or by hydrogasification–methanation (*see also* Chapter 6) and multiple methanation of lean gases produced by the oxidation or hydrogenation of coal (Chapter 9).

10.2 METHANATION CATALYSTS

The first successful reaction to produce methane from carbon monoxide and hydrogen was reported by Sabatier and Sanderens who used Raney nickel to effect the atmospheric pressure combination between the two gases at 280 to 290 °C. While nickel as the active component in methanation catalysts has been used ever since, there have been numerous attempts to substitute other metals that would be less affected by sulphur and other impurities in the feed. On the basis of the work of Fisher and Tropsch in Germany, cobalt deposited on kieselguhr was found to be effective, particularly if promoted by small concentrations of thoria or magnesia which were used as carriers for the active metals.

A simple attempt was made to reduce the sulphur sensitivity of nickel catalysts by means of co-catalysts or catalyst carriers that would delay or inhibit the deactivation of the catalyst. This class of catalyst includes one developed by the British Gas Corporation, a mixed nickel and aluminium deposit on a support material of china clay. Although higher reaction temperatures had to be used, the rate of carbon deposition which at high temperature becomes a problem could be restricted to relatively low rates at temperatures no higher than 350–400 °C inlet temperature. The further addition of manganese to the nickel–alumina on china clay catalyst was found to further reduce the sensitivity of the catalyst to sulphur compounds and it became possible with such a catalyst to methanate Lurgi coal gas without incurring too rapid a deactivation.

The alternative to developing sulphur-resistant catalysts is to purify the gases due to be methanated to an extremely low sulphur content. If this is done, and particularly if one makes sure that carbonyl sulphide is absent, the use of high nickel content catalysts without the addition of further co-catalyst becomes possible. Under those circumstances for instance one could use a low temperature reforming catalyst such as the British Gas Corporation's CRG catalyst or the 'BASF recatro' catalyst to convert gases of the type produced in low temperature reformers by a further processing stage at a lower temperature. In practice this means that a second, sometimes identical, reactor is added to a low temperature reforming train and that the gases that have been precooled down to 350 °C are passed over the catalyst bed.

It must, however, be realised that carbon monoxide present in the methanation feedstock will tend to react with the nickel catalyst and form nickel carbonyl. This highly volatile material is also extremely toxic and its formation must be prevented in order to maintain the catalyst in its active state and also to avoid accidents due to gas leakage. The formation of nickel carbonyl could be prevented by both temperature rise and pressure reduction. Since the latter is not practicable, methanation by means of a nickel catalyst imposes both a lower temperature limit on the process (to prevent carbonyl formation) and an upper limit, since methanation is exothermic and the resultant temperature increase could both deactivate the catalyst and shift the equilibrium in the reverse direction, *i.e.* towards carbon monoxide and hydrogen.

Another point to be remembered in the study of methanation catalysts is their relative activity in regard to carbon monoxide and carbon dioxide. Some catalysts will produce complete methanation of carbon monoxide long before the dioxide has been effectively converted. Other catalysts exert a balanced action on the two gases, and complete removal of carbon monoxide will require the simultaneous conversion of all the CO_2 present in the gas. Clearly, where there is a surplus of CO_2 present in the gas after the shift reaction, it will be necessary to remove some of it by alkaline absorption. Under those circumstances, a catalyst that converts all the carbon monoxide without complete carbon dioxide methanation will be preferred. If, on the other hand, the gas to be converted contains mainly hydrogen and only small quantities of carbon oxides, the second type of catalyst which converts all carbon oxides at more or less the same rate will be selected.

A further important point to remember is the sensitivity of some methanation catalysts to the presence of steam. Whilst properly prereduced metallic catalysts will generally accept the presence of up to 50 % H_2O without premature deactivation, a catalyst that has been reduced with hydrogen and carbon monoxide that itself has not been carefully dried, will generally be much more steam sensitive.

It is always necessary to prereduce freshly charged catalysts since the active metallic components are only effective in the reduced form and not as oxides. Reduction of a catalyst is considerably easier at higher temperature and the fact that most methanators are designed for heat exchange between exit and incoming gases in order to maintain methanation temperature constant does not help in the prereduction step. Here it will be desirable for the gas to enter the methanator at a temperature of at least 400 °C and it may be necessary to provide special preheat facilities and a bypass line for such a reducing gas.

The high degree of exothermicity of the methanation reaction and the fact that at high temperatures the methanation equilibrium shifts towards the carbon monoxide/hydrogen side makes it necessary to cool the reaction either internally or externally.[1] In the past, methanation of small concentrations of either hydrogen or carbon monoxide resulted in small temperature rises that could be accommodated by allowing the temperature of the diluent gas to rise. Once the concentration of the reacting gases becomes higher it is necessary to introduce intermediate cooling stages or to resort to other means of internal cooling. Thus proposals have been made recently to carry out methanation in an inert liquid that is simultaneously cooled to maintain constant temperature. Organic fluids, usually aromatic hydrocarbons, have been used, their boiling point depending on the process pressure since one must make sure that little or no solvent is lost by evaporation.[4] An alternative approach to the constant temperature methanation of highly reactive gases is the use of a fluid bed of catalyst, which again permits simultaneous reaction and cooling of the catalyst and the reacting gases.[3]   Both liquid-phase and fluid-bed methanations have certain disadvantages such as the inevitable back-mixing that prevents complete conversion of the reacting gases. It is therefore common practice to follow up fluid-bed or liquid-phase processing by a fixed-bed catalytic conversion.

10.3   METHANATION OF LOW TEMPERATURE REFORMER GAS

In a low temperature reforming process, naphtha or other light hydrocarbons are passed over a high nickel content catalyst at a temperature of 450 to 500 °C and undergo a reaction with steam that produces a mixture of around 21 % vol carbon dioxide, 25 % vol hydrogen and nearly 53 % vol methane with minor concentrations of carbon monoxide (all percentages calculated on a dry gas basis). Since, however, only part of the steam added to the naphtha in the first reforming stage reacts, the gas will still be diluted in a ratio of about 1:1 with unreacted steam. The gas leaves the reforming reactor at a temperature of about 550 °C and consequently has to be cooled before undergoing a methanation reaction.

The latter can be initiated at 250 to 300 °C, and low temperature reformers invariably incorporate gas cooling upstream of the first methanation stage. Cooling the gas also reduces its water content and a typical steam concentration at the entrance to the first methanation stage will be between 55 and 60 %. This corresponds to a temperature reduction to about 40 °C, water removal at that temperature and reheating of the gas, generally by heat exchange with methanator exit gas, to nearly 300 °C.

The gas now enters the methanation reactor and equilibrium between hydrogen and carbon oxides will soon be established. However, this results in a rapid temperature rise and, generally speaking, no more than about half to two-thirds of the hydrogen content can be allowed to react with the carbon monoxide and carbon dioxide in the gas. This produces a temperature rise of about 70 °C and the corresponding equilibrium concentration of methane is clearly lower than it would be if the entry temperature could be maintained.

TABLE 10.2
Methanation of Low Temperature Reformer Gas
(ex Japan Gasoline Co., MRG Reformer)

Composition mol %	Inlet to 1st stage	Outlet from 1st stage	Inlet to 2nd stage	Outlet from 2nd stage
CH_4	22·42	26·08	49·60	50·18
H_2	10·80	3·78	2·67	0·61
CO	0·29	0·08	0·06	0·01
CO_2	9·04	7·86	14·74	14·35
Steam	57·45	62·20	32·93	34·85
Combustion characteristics				
dry: Heating value, Btu/ft³	627·1	720·6	—	752·1
Relative density, 15°C, 1 atm (air = 1)	0·649 5	0·708 1	—	0·747 1
Wobbe Index	778·2	856·3	—	870·2
Weaver flame speed factor	—	18·2	—	15·2
dry, CO_2 reduced[a]				
Heating value, Btu/ft³	—	900·75	—	954·0
Relative density, 15°C, 1 atm (air = 1)	—	0·505 1	—	0·540 4
Wobbe Index	—	1 267·4	—	1 297·7
Weaver flame speed factor	—	18·4	—	15·3
Operating conditions				
Typical				
Temperature, °C	286	263	260	278
Pressures, atm[b]	9·6	9·4	15·6	15·4

[a] CO_2 reduced to 1·0% vol.
[b] This is a low pressure city gas operation. For pipeline gas, operating pressures of about 25 to 40 atm would be used.

It is therefore desirable to carry out the methanation reaction in several steps, usually two (*see* Fig. 6.2), and if one reduces residual hydrogen to about 2% vol in a previous stage, it becomes possible to carry out the last stage in the absence of steam. Otherwise, *i.e.* with 3% hydrogen or more in the exit gas from the first methanator, it is desirable to dilute the gas with additional steam, or in other words, not to remove all the residual water, in order to prevent an excessive temperature rise in the last stage.

Table 10.2 shows gas compositions at the inlet to the first stage, outlet from the first stage, inlet to the second stage, and outlet from the second stage of a methanator used in conjunction with a typical low temperature reforming plant (MRG process). Combustion character-istics shown in the table do not refer to the gas composition as it stands, but to the gas either after it has been dried or after it has been dried and its carbon dioxide content reduced to 1.0% vol. The first methanation stage thus produces a gas containing about 10% vol hydrogen (on a dry basis) and having a Weaver flame speed factor of 18.4. As discussed in Chapter 3, this would be marginally acceptable in a US appliance population but provides little or no room for error. As was also mentioned in Chapter 6 in respect of the CRG process, in order to produce an interchangeable SNG it is preferable to methanate in at least two stages; it is, however, permissible to have steam present in both stages, in the first to the extent of over 60%, in the second stage at about 30% of the total gas stream.[2]

TABLE 10.3
Final Methanation of Gasynthan Naphtha Gas

	After wet methanation	After dry methanation
Composition, mol %		
dry, CO_2 reduced		
H_2	1·66	0·72
CO	0·02	0·10
CO_2	1·00	1·00
CH_4	97·32	98·18
Combustion characteristics		
Heating value, Btu/ft^3	992·3	999·2
Relative density, 15 °C		
1 atm (air = 1)	0·556	0·560
Wobbe Index	1 331·0	1 335·2
Weaver flame speed factor	14·4	14·2
Operating conditions		
Outlet temperature, °C	310	357
Pressure, atm	42·7	42·8

Since it is in fact possible to operate in the absence of steam in the final methanation stage, one can produce, provided the same exit temperature can be maintained, an even higher methane concentration than in the case of wet methanation.

This is demonstrated in Table 10.3 which compares the result of methanating Gasynthan gas containing about 5% vol hydrogen, first in the presence of 35% vol steam, then under dry conditions.

10.4   METHANATION OF HYDROGASIFICATION PRODUCTS

The hydrogasification of a petroleum feedstock can be used in at least two of the many possible routes to SNG; one can, for example, hydrogasify naphtha using hydrogen formed in the course of the low temperature reforming of a light feedstock. This means that naphtha or a similar hydrocarbon is reacted at a temperature of 450–500 °C with steam over the usual reforming catalyst, as discussed in Chapter 6, and that the gas produced in this way is further reacted with additional feedstock over a similar catalyst, but at a somewhat lower temperature, say between 400 and 450 °C. Residual steam in the gas, hydrogen and fresh feed react to produce a gas containing about 6 to 8 % vol of residual hydrogen, compared with about 20 % before hydrogasification, and also resulting in a somewhat increased methane content.

Alternatively, naphtha or a light distillate or even gas oil can be gasified by means of hydrogen gas in a Gas Recycle Hydrogenator as described in Chapter 7. Hydrogen for the process is generated externally, either from by-products (aromatics) or from recycled product gas. However, at least some of the hydrogen does not react in the recycle reactor and finds its way into the GRH product gas. In order to optimise its utilisation it then becomes mandatory to use the GRH gas, either exclusively or as additional feedstock to the first or second stage of a catalytic low temperature reformer. Under those circumstances the bulk of the hydrogen reacts, and the effluent gas, as shown in Table 10.4, again contains only about 8 % vol of residual hydrogen (*see* columns 1 and 3).

In both instances, this residual hydrogen has to be removed by a final methanation step and as can be seen from columns 2 and 4 of Table 10.4, the result is SNG of satisfactory quality (97·5 % vol CH_4). In neither case is it necessary to dilute the methanation feed with steam as it is generally found advantageous to methanate such a gas under dry conditions.

A point to bear in mind in connection with the processing of GRH gas is that the latter is not normally sulphur free. It will, therefore, be necessary to remove sulphur compounds by means of a desulphurisation catalyst/sulphur adsorbent combination prior to contacting the methanation catalyst. However, this already applies to the second reforming (hydrogasification) stage of the process and

TABLE 10.4
Methanation of Hydrogasification Products

	CRG Hydrogasification		SNG Production via GRH	
	Before methanation	After methanation	Before methanation	After methanation
Gas composition				
mol % dry				
CH_4	70·93	97·61	72·95	97·8
H_2	6·54	0·80	7·95	0·6
CO	0·67	0·09	0·60	0·1
CO_2	21·86	1·50	18·50	1·5
Gas characteristics				
Heating value, Btu/ft³	—	1 000	—	1 001
Relative density,				
15°C, 1 atm				
(air = 1)	—	0·555	—	0·555
Wobbe Index	—	1 342	—	1 343
Weaver flame speed factor	—	14·2	—	14·2
Methanator operating conditions				
Inlet temperature, °C	280		265	
Pressure, atm	34·0		40·0	
Steam	absent		absent	

does not impose any additional burdens on the final methanation step.

Carbon dioxide removal is required in most instances in order to raise the calorific value of the residual gas to that of natural gas and so as to match its other combustion characteristics. Values for the latter quoted in the table, therefore, refer to the final fully purified gas.

10.5 METHANATION OF LEAN GASES[5]

In order to convert coal or residual fuel into a gaseous fuel, as discussed in Chapter 9, it is generally necessary to employ hydrogen or oxygen at some stage. The lean or intermediate gas produced in such a process tends to have a high content of hydrogen and carbon oxides. Having adjusted the H_2/CO ratio appropriately by a water gas shift reaction, one invariably ends up with a highly reactive gas that cannot be adiabatically methanated since temperature rise would be excessive and the final temperature unfavourable for the methane equilibrium.

In order to arrive at a reasonable outlet temperature, *i.e.* below 450 °C, multi-step methanation with intermediate cooling, or better still, some form of internal cooling of the reaction zone, becomes essential.

Possibilities for limiting the temperature rise in an exothermic reaction are numerous: they include dilution with inert gas or cooling of recycled product gas; introduction of indirect coolers into the reaction space; stepwise addition of cooled reactants; operation in the liquid phase in a solvent which can be kept at constant temperature by evaporation; reaction in a fluid bed of methanation catalyst, etc. All present certain problems, mainly of economics, and at the moment it is not easy to say which will eventually be adopted. It is, therefore, proposed to discuss three fundamentally different approaches in somewhat greater detail.

10.5.1 The IGT Cold Feed Quench
This type of methanation has been proposed and applied on a pilot scale to gas produced by hydrogasification of prepared coal in a series of pressurised superimposed fluidised beds in the HYGAS process, but should also be applicable in other situations where gases of similar composition are produced.

The gas enters the methanation section of the plant after removal of sulphur compounds and other contaminants and cooling to condense excess steam, the water having been drawn off, at a pressure of 70 atm and a temperature of 38 °C. The methanation plant consists of three stages: each reactor has a top inlet designed to take about 10 % of the feed and three side inlets used for quenching by means of the rest. Only the 10 % of the feed that is to initiate the reaction is preheated by heat exchange with the outgoing gas to the reaction temperature—about 260 °C—and the remaining 90 % is injected cold.

Under these very carefully controlled conditions only partial methanation takes place, but even so the gas temperature rises to about 480 °C. Chemical equilibrium between the different species is not attained, however, and the effluent gas is again cooled to 38 °C and put through a second methanation stage that closely resembles the first. Again, preheated gas is injected through the top inlet and quench streams enter through the side; again incompletely equilibrated gas leaves the bottom of the reactor at 480 °C and is cooled to room temperature before entering the third and final stage.

At that point, hydrogen and carbon oxides have been reduced sufficiently to limit the temperature rise to 414 °C, *i.e.* the effluent gas composition now corresponds to the thermodynamic equilibrium at that temperature. Reduction of residual carbon dioxide to 1·0%, cooling of the gas and final separation produce a product that is fully interchangeable with standard quality natural gas.

10.5.2 The BI-GAS Fluid-Bed Methanator[3]

The BI-GAS fluid-bed gasifier is designed to produce a gas of extremely high content of reactive components, *i.e.* 21% vol carbon monoxide, 63% vol hydrogen and only 16% vol methane obtained by steam–oxygen pressure gasification of bituminous coal. The massive heat evolution produced by reacting the gas over a methanation catalyst would at first appear to be uncontrollable and a single-stage methanation originally seemed out of the question.

The main modifications of the classical methanation route adopted in the BI-GAS process were an increase in reaction temperature and a change of catalyst. The BI-GAS methanator operates at a controlled and uniform temperature of 430 °C and being a fluid-bed reactor the gas inlet temperature is in fact the same as the reaction temperature. A particularly robust catalyst of low activity is required in place of the highly active nickel catalysts used in other systems.

The higher operating temperature has advantages in regard to heat recovery and process economics. The fluid bed is cooled internally by passing either Dowtherm fluid (in the pilot plant) or boiler feed water under pressure (in the full scale plant) through finned heat exchanger tubes, the high pressure steam or heat exchange fluid leaving the reactor at over 300 °C. The well stirred fluidised bed ensures uniform temperature without local hot spots and intensive heat exchange between the cooling tubes and the bed. However, it will be appreciated, that owing to back mixing, conversion of carbon monoxide and hydrogen to methane in a BI-GAS methanator can never be complete. In fact, in all the full scale BI-GAS coal gasifiers, the fluid-bed methanator will be followed by a cooler, water removal and a final stage of fixed-bed methanation designed to lower carbon oxide content to about 0·8% vol and to raise methane content to above 95% vol.

No final decision has been taken whether the BI-GAS fluid-bed methanation system will be run as a once-through or recycle unit. However, even a 2:1 recycle to fresh feed ratio would still result in a residual hydrogen content of more than 8% and would not altogether eliminate the final fixed-bed stage.

10.5.3 Liquid Phase Methanation[4]

This process also is intended to limit the temperature rise consequent upon reacting high concentrations of hydrogen and carbon monoxide in an adiabatic reactor. The heat of reaction, in this instance, is removed in part as sensible heat, in part as heat of evaporation of a circulating inert solvent. Heat recovery is sufficiently high to provide high pressure steam for the other stages of the process.

The liquid phase not only removes excess exothermic heat from the system but also, owing to its circulation, assists in fluidising the catalyst, a standard nickel material on an inert support, *e.g.* CIC C–150–1–02. However, the main advantage of liquid-phase operation over fluid-bed gaseous-phase methanation is the much reduced degree of back mixing. The reacting gases travel very nearly in plug flow through the reactor and the exit gas is much closer to thermodynamic equilibrium than in the case of fluidised-bed methanation.

Operating conditions in a liquid-phase methanation pilot plant are approximately as follows: operating pressure is 75 atm, inlet gas temperature 38 °C, liquid-phase reactor outlet temperature 300 °C, condenser and solvent knock-out drum temperature 40 °C. Condensed solvent is returned to the reactor together with fresh feed gas and recycle gas, if required. Catalyst can be withdrawn and reduced in a separate reactor at up to 400 °C. Solvent used in the process is a mixture of C_9 to C_{11} aromatic liquids, which was found to have suitable boiling characteristics under these conditions. Clearly the solvent must also be separated fairly easily from entrained process water and carry-over catalyst.

Effluent gas after cooling and solvent and water removal can undergo a final fixed-bed methanation step that reduces both hydrogen and carbon oxides to full natural gas interchangeability. Even allowing for such a final stage, the designers of the process, Chem. Systems Inc., claim that it is substantially more economic for the finishing of gases derived from coal or residual fuel, than are competing processes such as multiple quench systems or fluid-bed methanators.

10.6 CONCLUSIONS

It can be concluded that methanation, both as a final trim of gas quality and as an essential processing step in the conversion of liquid and solid fuels into SNG, is of considerable significance. Being an exothermic process, it presents problems such as control of operating temperature, and the economic utilisation of the process heat generated in this stage of conversion.

There are a number of approaches, some tentative and still under development such as fluid-bed or liquid-phase methanation; others reasonably well established such as wet methanation and multi-stage methanation are an integral part of complete commercial scale systems for the manufacture of SNG. A need for conversion processes tailor made to specific raw gas compositions and well defined gasification process operating conditions is indicated.

REFERENCES

1. Allen, D. W. and Yen, W. H. (1973). Methanator design and operation, *Chem. Engng Prog.*, 75–79.
2. Fukunaga, K., Nojima, S., Okagami, A. and Ward, D. J. (1973). *Methanation Experience at Keiyo Gas*, 5th Pipeline Gas Symposium, Chicago.
3. Grabeski, M. S. and Diehl, E. K. (1973). *Design and Operation of the BCR Fluidised Bed Methanation PEDU*, 5th Pipeline Gas Symposium, Chicago.
4. Sherwin, Martin B., Frank, Marshall E. and Blum, David B. (1973). *Recent Developments in Liquid Phase Methanation*, 5th Pipeline Gas Symposium, Chicago.
5. Lee, A. L. (1973). *Methanation for Coal Gasification*, Paper 18, Inst. Gas Tech., Clean Fuels from Coal Symposium.

Chapter 11

The Economics of SNG Production

11.1 INTRODUCTION

The economics of the majority of industrial conversion processes have a number of aspects in common: one is concerned with one or more raw materials, their rate of consumption and their delivered cost; one incurs certain conversion expenses, usually due to the cost of labour, supervision, utilities, auxiliary materials, plant maintenance, purchased outside services and overheads; next there are financial charges such as interest on construction loans and a minimum return on promoters' funds; a depreciation provision ensures that loans and other funds provided are reimbursed by the time plant operations cease; finally one has to allow for some profit to be made by the operator and tax may be payable on such income. All these costs have to be met by sales of the product or products, whose minimum ex-plant price must be such as to match these requirements. A somewhat more detailed discussion of the usual costing of chemical conversion processes will be found in Appendix C.

While all these considerations apply to the conversion of liquid and solid hydrocarbon feedstocks into SNG, there are certain simplifying assumptions that can be made in such a system. In particular it will be found that most major SNG plants will be autonomous in regard to steam and power, i.e. consumption will be sufficient to justify generation within the plant. The number of input items is thus reduced; usually there is only one feedstock, and auxiliary materials are mainly boiler feed and cooling water in addition to various chemicals and catalysts.

Products of these plants are SNG, sulphur and sometimes desulphurised middle distillates and coke. Intermediate products such as oxygen, hydrogen, steam, electric power, fluid coke, aromatics etc. are consumed at the rate at which they are formed, if this is at all possible, and can therefore be disregarded.

A concept of great utility is the gasification yield. If only one feedstock is converted and is also used to fire boiler plant and to generate power, and if the only product is SNG, then the total heat content of the product will amount to a certain proportion of the thermal input; this ratio is known as the thermal yield and is identical, in these circumstances, with the gasification yield. If, on the other hand, feed and fuel are separate and distinct or if fuels other than SNG are produced in substantial quantities then thermal and gasification yields will clearly no longer be the same.

In recent years throughput capacity of SNG plants has risen to very high levels; plants of up to 500 million cubic feet per day are under construction in the US. Since investment is, of course, not directly proportional to plant size, and large plants require no more operating labour than small plants, provided they have the same number of trains, the conversion cost in such units is much reduced. Feedstock cost, on the other hand, is constant for plants of the same thermal and gasification efficiency. As a result the split between feed and conversion costs tends to shift towards the former and in very large plants the cost of conversion can become insignificant. The cost of SNG then amounts to little more than feedstock cost divided by gasification efficiency.

This assumes, of course, that these very large SNG plants operate continuously and that their annual send-out volume is close to their capacity. SNG plants that operate only part-time to meet peaks of gas demand will not only produce gas at a higher basic cost, largely because of increased capital charges, but their operating cost then becomes significant compared with total gas cost.

In order to compare SNG costs for different manufacturing routes, it is desirable to discount, as far as possible, feedstock prices. The reasons for doing so, apart from the different order of magnitude of feedstock and operating costs are the following:

prices of petroleum products vary over a fairly wide range,
there are considerable geographical differences due to variations in product mix in different markets,
there have been many gradual shifts and a few very sudden changes in petroleum prices in the past.

For all these reasons it makes sense to calculate the cost of converting the raw material into SNG, sometimes also called the gas service cost, rather than the total cost of operating the various SNG processes that were reviewed in previous chapters. Service cost has therefore been made the standard on the basis of which SNG processes are compared, and feedstock cost is subsequently introduced as a variable.

Service cost can be expressed either as the total operating cost, less the cost of feedstock, per operating day (M$/day etc)* or in terms of an operating cost per unit thermal send-out (¢/million Btu, pence/therm,

* M in this text equals 1000; MM equals 10^6.

TABLE 11.1
Basis of Calculating SNG Manufacturing Cost

(1) Scale of operation:
    250×10^6 Scfd of SNG (gross cal. value 1000 Btu/Scf)

(2) Load factor:
    90%; equal to 330 days' operation per year (approximately)

(3) Capital cost basis:
    Location and year of construction: Japan mid 1975
    Battery limits (total inside plot) investment
       includes process units, control room, instrumentation, etc.
    Total capital investment equals 130% of on-site investment,
    includes feedstock reception and storage, roads, offices, stores,
    fences, metering, etc.

(4) Annual operating cost:
    (a) Fixed charges:
        Capital charges: 20% of total investment, to cover repayment of capital, return, tax, etc.
        Operating costs: 6% of total investment, to cover labour, supervision, maintenance, overheads

    (b) Variable charges:
        Catalyst and chemicals; By-product credits
        Utilities (ex fuel); Fuel

Prices:		
Electricity	$ 15·0	per 1000 kWh
LP steam (3–5 atm)	0·7	per metric ton
MP steam (18 atm)	1·3	per metric ton
HP steam (40–50 atm)	1·4	per metric ton
Cooling water (circul. system)	1·5	per 1000 m³
Boiler feed water	350·0	per 1000 m³
Sulphur (credit)	35·0	per metric ton
Aromatic Liquids ⎱ Coke ⎰	Fuel value	

(After Ref. 7.)

DM/Gcal etc). The latter form will be found most useful, since final cost of the SNG produced simply reduces to service cost plus feedstock cost provided both are expressed in the same units.

In order to compare effectively different SNG processes it is furthermore necessary to use standard values for capital cost and service factors, financial charges, price of utilities, labour, chemicals and by-products. Table 11.1 lists the assumptions made throughout this book in respect of the above. The same assumptions are made in the calculations of gas service cost as explained in Appendix D.

11.2   ECONOMICS OF NAPHTHA GASIFICATION[7,9]

The standard method of manufacturing SNG from a naphtha or LPG feedstock is that set out in Chapter 6, *i.e.* low temperature steam reforming. The steps involved are: vaporisation of the feed, hydrodesulphurisation with recycled hydrogen over a nickel or cobalt molybdate catalyst, steam reforming of desulphurised hydrocarbons over a high nickel content catalyst, two-stage methanation, removal of surplus carbon dioxide, gas cooling and gas drying. In addition some of the steam reformer exit gas is recycled, subjected to high temperature reforming followed by a carbon monoxide shift and CO_2 reduction, and the resultant hydrogen gas is blended with the fresh feedstock ahead of the feedstock desulphurisation section. In the Catalytic Rich Gas and the Gasynthan processes the first of the methanation stages can be replaced by a second, lower temperature, steam reforming (or hydrogasification) stage, in the course of which surplus steam and hydrogen react with additional hydrocarbon feed which is injected upstream of the second reformer and is hydrogasified in the latter.

The cost of service of the CRG, of the Gasynthan and of the Japan Gasoline Co.'s MRG processes is shown in Table 11.2. Estimates of investment costs are also given and these of course are reflected in the fixed charge element of the operating costs in every case. It must be said that since these numbers were first derived at the final drafting of this book, all cost estimates have become dated very quickly, due to the incidence of very high inflation rates. However, it is considered that comparative process economics are still valid since the data all relate to the same period in time. So, though investment estimates may now be on the low side, particularly those of coal based processes, the conclusions drawn as to which of the processes discussed in this chapter are the most expensive, which least, etc., are probably not in dispute with the proviso that it is always more problematic to compare costs of processes still in development to those of developed processes since the former are more subject to escalation for technical reasons than the latter.

TABLE 11.2
Economics of Naphtha Low Temperature Steam Reforming Processes

	Catalytic Rich Gas with hydro-gasification	Gasynthan with 2-stage gasification	Japan Gasoline Co. MRG with methanation
Investment, MM$			
On site	60·8	65·4	61·5
Off site	18·2	19·6	18·5
Total	79·0	85·0	80·0
Cost of service, $/sd			
Capital charges	47 900	51 500	48 500
Operating, maintenance, overheads	14 400	15 400	14 500
Total fixed charges	62 300	66 900	63 000
Catalysts and chemicals	12 500	11 000	13 000
Electricity	3 000	3 000	3 500
Make-up water, etc.	13 000	13 000	13 000
Total variable charges	28 500	27 000	29 500
Total cost of service,			
—$/d	90 800	93 900	92 500
—¢/MMBtu	36·3	37·6	37·0

The table shows the relatively small differences in investment between the three processes and the even smaller variations in the cost of gas service, once artificial differences such as inconsistent assumptions in regard to project definition, financial charges, labour and utility costs have been eliminated.

It is worth noting that the naphtha feedstock cost which at the time of writing might approximate to about 285–300 ¢ (range 270–380) per MMBtu corresponds to about 80 % of total gas cost whereas cost of service amounts to only 20 % (v. Fig. 11.1). It is also of interest that choice of process and the relative gasification efficiencies of the different processes have little or no effect on gas cost. And while investment savings of up to 7 % or $6 million for a 250 million ft^3/d plant are no doubt worth having, the effect of such a saving could easily be swamped by higher catalyst costs or a lower gasification efficiency.

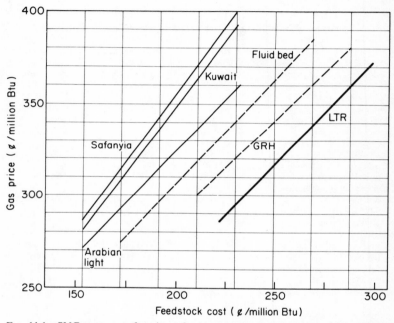

FIG. 11.1   SNG cost as a function of process route and feedstock price—liquid hydrocarbons. — energy refinery; --- hydrogasification; ▬ low temperature reforming.

On the other hand it should be borne in mind that for a plant that operates fewer hours per year these relatively small changes in operating cost may become significant.

The alternative route to SNG from light hydrocarbon feedstocks involves a Gas Recycle Hydrogenator as a first step in the conversion. The system, discussed in more detail in Chapter 7, is slightly more complex than low temperature reforming. It consists of GRH, hydrodesulphurisation, heavy aromatics removal, low temperature steam reforming and methanation, with recycle hydrogen for the GRH and desulphuriser produced from the LTR exit gas by high temperature reforming, CO conversion and CO_2 removal. But its obvious advantage over low temperature reformers of the CRG type is greater feedstock flexibility. Hence wherever it is necessary to gasify middle distillates up to an end point of 340 °C the gas recycle hydrogenator route will be considered and the economics set out in Table 11.3 will apply.

TABLE 11.3
Economics of Hydrogasifying Naphtha Feeds and Light Distillates

Feedstock...	Naphtha	Kerosine/Gas Oil
Investment, MM%		
On site	83·7	88·5
Off site	25·0	26·5
Total	108·7	115·0
Cost of service $/sd		
Capital charges	65 900	69 700
Operating, maintenance, overheads	19 800	20 900
Total fixed charges	85 700	90 600
Catalyst and chemicals	10 500	13 200
Electricity	3 200	3 200
Make-up water	12 500	12 500
Aromatics disposal[a]	4 800	12 000
Total variable charges	31 000	40 900
Total cost of service		
—$/d	116 700	131 500
—¢/MMBtu	46·7	52·6

[a] Feedstock minus fuel $2·0 per brl, *e.g.* feedstock at $10 per brl, fuel at $8.

Table 11.3 shows that middle distillate processing requires a somewhat higher investment than does naphtha conversion. Its capital charges are accordingly higher. Furthermore the higher sulphur content and lower volatility of the feedstock increase the consumption of catalysts and chemicals. Finally the higher yield of liquid aromatics increases the cost of their disposal.

It should be pointed out, however, that the calculation of a cost of service in this instance is no longer absolutely clear-cut; while aromatic by-products of the gasification of a naphtha feedstock are of the order of 10% and, therefore, can be easily consumed as process fuel, this is no longer the case if kerosine or gas oil is hydrogasified. In the latter case up to 25 % of the thermal content of the feed finds its way into the by-products and since it is unlikely that these materials could be marketed as chemicals appropriate allowances have to be made for their disposal.

In the second column of Table 11.3, for example, it has been assumed that by-products of a gas oil hydrogasification facility could be disposed of as fuel, *e.g.* to a nearby steam power plant, and that they would be priced 10 % below the cost of the hydrocarbon feedstock. It must be admitted, however, that this hypothesis may not always be valid and that our conclusions in this instance are not as generally applicable as we would wish.

A comparison of the service cost of naphtha gasification in Table 11.2 of about 36 ¢/million Btu with that of middle distillate conversion of about 50 ¢/million Btu indicates that only a correspondingly lower feedstock cost would allow hydrogasifiers to compete effectively with low temperature reformers. The value, as far as SNG production is concerned of a middle distillate can thus be said to be about 14 ¢/million Btu or roughly $1·0 per barrel less than that of naphtha.

11.3 ECONOMICS OF CRUDE OIL GASIFICATION[1,2,5,8]

The relatively simple gasification systems that can be used for the conversion of light hydrocarbon fuels such as naphtha or LPG are no longer sufficient if it is desired to convert the bulk of a crude oil, including its residual fractions, into SNG. As discussed in Chapter 8, it is generally desirable under those circumstances to treat light and heavy fractions separately, to shift middle distillates into the range of, and conversion systems for, light petroleum products by some form of hydrocracking, to use hydrogen, often prepared by partial oxidation of heavy products with oxygen, for desulphurisation, hydro-gasification and hydrocracking, or to remove carbon from the system in the form of solid coke. Final products can be just SNG, or SNG and various desulphurised liquid fractions, or SNG, liquid fractions and coke. A further potential variable is the feedstock which can range from very light low sulphur crudes such as Libyan Sarir, or Algerian Hassi Messaoud to heavy Arabian or Khafji which are high in sulphur, or to Venezuelan Bachaquero, which is made up mainly of heavy asphaltenic hydrocarbons.

To make unqualified statements about the cost of manufacturing substitute natural gas from crude oil would, therefore, be totally unjustifiable. It would similarly be very difficult to generalise about the cost of service in different refineries or conversion systems. Even to compare the cost of gas from different feedstocks will be difficult since, clearly, one would have to optimise processing route, processing conditions and by-product yield for each competing raw material before one could hope to compare them on a strictly economic basis.

In the light of these complications, the only practicable approach seemed to be to select almost at random a number of feedstocks and process routes that had been examined and recommended for specific situations (no complete crude to SNG plant or energy refinery has so far been completed or even projected) rather than to analyse actual facilities and to report their economics. No attempt has been made here, in view of the above mentioned complexity, to ensure that the basis of each of the projects under review was strictly and absolutely comparable. On the other hand, plant capacities and locations, product quality and other more easily adjustable variables such as financial charges, maintenance expenditure and the cost of utilities have been standardised on the basis of the assumptions listed in Table 11.1.

The first column of Table 11.4 analyses an all-SNG case. Feedstock for the 250 million scf per day facility is Kuwait crude 59 000 bd of which will be consumed. The plant consists of units for primary distillation (topping), vacuum distillation, hydrocracking of atmospheric and vacuum gas oil, partial oxidation for the vacuum residue with its oxygen plant, hydrodesulphurisation and gasification for straight-run and hydrocracked naphtha, hydrogen purification and sulphur recovery.

In the second column a different series of processes has been considered. Again the aim of the conversion is a maximum yield of SNG and the net yield of by-products has been reduced to the lowest possible level. This has been achieved by processing the entire feed in a fluid-bed hydrogenator (FBH), described in Chapter 7, from which gas and liquid by-products and some coke are withdrawn; liquids are fed to a partial oxidation plant and yield hydrogen for FBH processing and desulphurisation; gaseous product is desulphurised and subjected to a second hydrogenation stage.

The data in the table refer to a light Algerian crude oil (Hassi Messaoud) and are, therefore, not strictly comparable with the

TABLE 11.4
Economics of Crude Oil Gasification to SNG

Crude	Kuwait	Algerian	Arabian Light
Gasification system	Partial oxidation/ hydrocracking	Partial oxidation, fluid-bed hydrogenation	FLEXI-COKER,[5] hydrocracking
Products	SNG, sulphur	SNG, sulphur, coke, liquids	SNG, sulphur
Investment, MM$			
On site	192·3	132·0	190·2
Off site	57·7	39·6	57·1
Total	250·0	171·6	247·3
Cost of service, $/sd			
Capital charges	151 500	104 000	149 800
Operating, overheads, maintenance, etc.	45 400	31 200	44 900
Total fixed charges	196 900	135 200	194 700
Catalysts and chemicals	17 000	15 000	12 000
Electricity	8 000 ⎫		
Boiler feed water, cooling water	21 000 ⎬	29 000	21 000
Sulphur credit	(7 000) ⎭	—	(5 000)
Coke and liquids credit	—	(1 000)	(1 000)
	39 000	43 000	27 000
Total cost of service:			
—$/sd	235 900	178 200	221 700
—¢/MMBtu	94·4	71·2	88·7

processing cost of Kuwait crude in the first configuration. However, it would seem that substantial economies both in investment and operating costs can be effected by switching from the first mentioned processing series to FBH.

Similarly, the refinery configuration involving FLEXICOKING, which is evaluated in the third column, appears to result in savings compared with the partial oxidation–hydrocracking route. though these are less than the economies estimated for FBH, etc. In the FLEXICOKING process, it may be recalled, crude oil is topped and vacuum distilled, the vacuum residue being transferred to the

FLEXICOKER. The latter produces coker naphtha, which is gasified, and a clean low heating value gas which can be used to release other refinery fuels for producing hydrogen, this being required in the gas oil hydrocracker to convert atmospheric and vacuum gas oil into naphtha and gas.

Again the operating cost figures are not strictly comparable since the FLEXICOKING scheme was designed and evaluated for light Arabian crude. However, it appears to produce SNG at only slightly higher cost than the FBH route, although the feedstock under consideration is heavier and higher in sulphur than that of the FBH scheme.

11.4   'ENERGY REFINERY' ECONOMICS[1,3,11,12]

The cost of converting crude oil and fuel oil into gaseous fuels is relatively high and involves the use of hydrogen in a processing scheme that can be easily modified to produce both low sulphur liquid fuels and SNG. An obvious method of reducing the service cost of the gas is, therefore, to credit the latter with the cost of desulphurisation of clean liquid fuels that are produced in parallel with the gas. The economics of SNG produced in an 'energy' refinery can thus be slightly more favourable than those of a plant for the exclusive manufacture of SNG.

Investment and operating costs in Table 11.5 refer to refineries of the same crude throughput as those of Table 11.4. Since part of the plant output is in the form of low sulphur fuel oil, SNG volume production from the energy refineries listed here is substantially lower than that of the SNG plants. However, cost of service figures expressed in terms of US cents per million Btu of SNG production are still comparable, due in two of the cases to reduced capital charges as well as to the sizeable desulphurisation credit.

The processing scheme in the first column of Table 11.5 is designed to convert Kuwait crude into low sulphur fuel oil (LSFO) and SNG in the proportions of 59 000 bd of feed to yield 32 300 bd of liquid products and 104 million scf per day of gas. The conversion facility consists of atmospheric distillation, desulphurisation of naphtha, gas oil and residue, hydrogen manufacture by HT reforming and naphtha gasification. That the cost of service under the circumstances is of the order of 70 ¢/MMBtu is mainly owing to the large price difference

TABLE 11.5
Economics of 'Energy' Refineries

Crude	Kuwait	Arabian Light	Arabian Light
Process	Gas oil and residue HDS, H_2 by HTR	Vacuum resid. FLEXICOKING, Gas oil HDS	Gas oil HDS Vac. residue partial oxidation
Products	SNG Sulphur, LSFO[a]	SNG Sulphur, LSFO	SNG Sulphur, LSFO
Investment, MM$			
On site	138·5	136·5	145·4
Off site	41·5	41·0	43·6
Total	180·0	177·5	189·0
Cost of service, $/sd			
Capital charges	109 000	107 720	114 500
Operating, maintenance, overheads...	32 700	32 200	34 300
Total fixed charges	141 700	139 920	148 800
Catalysts and chemicals	15 300	10 000	10 000
Electricity Make-up/cooling water, etc. }	7 000	8 000	18 000
Sulphur and coke credits	(6 000)	(4 800)	(3 800)
Desulphurisation credit[b]	(82 700)	(82 700)	(82 700)
Total variable charges	(66 400)	(69 500)	(58 500)
Total cost of service			
—$/d	75 300	70 420	90 300
—¢/MMBtu	72·4	67·7	87·0

[a] 104 MM scf/d SNG plus 32 200 bd low sulphur fuel oil.
[b] Low sulphur fuel oil has $2·6 per barrel, premium over feedstock.

(*e.g.* in Japan) between crude oil and desulphurised fuel oil, and the consequently high desulphurisation credit.

The second processing scheme under consideration is designed for light Arabian crude and consists of crude oil atmospheric fractionation and vacuum distillation and FLEXICOKING of the residue, desulphurisation of both naphtha and gas oil and gasification of the former, the gas oil being co-produced as a low sulphur liquid fuel; it also compares favourably with process schemes that do not produce LSFO.

The third scheme, again for light Arabian crude, differs from the former two in that hydrogen is produced from the vacuum residue by partial oxidation with oxygen. But, as distinct from the first two systems, it produces SNG at a slightly higher cost than the equivalent gasification plant designed exclusively for SNG production.

Any advantages this type of processing has over complete gasification will largely depend on the market premium of low sulphur fuels over crude oil. In the present context this has been assumed to be of the order of $2·6 per brl, but clearly this will depend on the availability of low sulphur crude or of desulphurisation facilities and demand. Particularly where public ordinances prohibit the use of high sulphur fuels, *e.g.* in the New York and Tokyo areas, the economics quoted above will apply; elsewhere this may not be the case.

11.5 THE ECONOMICS OF GASIFYING COAL[10]

While the demand for coal as a fuel in Europe combined with the cost of mining coal from thin and partly exhausted seams makes it unlikely that European coals would ever be widely used as a gasification feedstock, this is not the case in the United States, South Africa, Australia and some other countries. Particularly in the United States, high gas demand, the limited availability of light gasification feedstocks and the continuing short supply of crude petroleum have resulted in a great deal of interest in coal for gas making. A number of projects have progressed well beyond the initial planning stage and the cost of SNG from this source can by now be predicted with reasonable accuracy.

The main difference between coal and oil gasification economics is the effect of gasification plant location; while it is generally cheaper to ship oil to an SNG plant located in a gas-consuming area, coal, on the other hand, will more often than not be gasified at or near the coal mine. The reasons for this difference are, of course, the relative costs of transporting energy in the form of coal, gas and oil over long distances. Since in terms of both operating cost and investment at to-day's prices it is as a rule cheapest to move oil by pipeline, less cheap to conduct gas and most expensive to transport coal, it makes sense to gasify the latter before shipment and the former after shipment.

Again this preference is more obvious in the US where most cheap coals are produced by open-cast mining in the States of New Mexico and Colorado. Since a large number of gas pipelines are already in existence for the transport of natural gas from fields mainly in Texas towards the Eastern Seaboard little additional pipeline investment will be required to link future gas-from-coal plants to the natural gas transport system.

While most of the coal gasification processes discussed in Chapter 9 have been costed, the reliability of such cost estimates will clearly be proportional to the extent of technical and commercial development that has already taken place. Uniquely among the processes considered for large-scale manufacture the Lurgi process has already been used commercially in many parts of the world; Lurgi cost estimates can, therefore, be considered the most accurate among those quoted. Other cost estimates such as those for the HYGAS, the BI-GAS, the CO_2 Acceptor, the Synthane and the Union Carbide–Battelle processes, which have been tested in large pilot plants, are of a lower degree of accuracy, but should be looked at in the context of possible future plant construction. Finally, cost estimates for processes that, at the time of writing, have only operated on a bench scale must, at the moment anyway, be taken with some reserve.

An important feature of coal gasification technology is the width of applicability of the process in question. While some processes have been developed with a specific type of coal in mind, and their feedstock limitations are therefore of little consequence, other routes that are claimed to be more universal must be carefully examined in this respect. In some ways this makes it extremely difficult to compare SNG-from-coal economics in general terms; what one should do is evaluate situations, starting with cost and availability of the coal and proceed through the possible processing routes to the type and quality of the final gas required, rather than compare processes in their own right. However, since this would require time and space much in excess of that available here, there is practically no alternative to evaluating the economics of the different routes as such, at this stage, in spite of the above reservations.

The economics of the first two US Lurgi plants[10], which will both be located in New Mexico, are determined by their size; a send-out capacity of 250 million sfd will require the processing of over 20 000 tons of coal per day in a total of 30 Lurgi pressure reactors. Daily steam consumption will also be over 20 000 tons and 5 500 tons/day of oxygen will be needed. Economics as forecast in mid-1973 for the first full year of operation in 1977 are summarised in Table 11.6.

TABLE 11.6
Economics of Coal Gasification—Lurgi SNG Plant

Investment		
million $	Process units incl. 30 gasifiers	129·3
	Utility units	84·9
	Off sites	52·0
	Administration + general	1·2
	Catalysts, lubricants, chemical	5·7
	Sub-total	273·1
	Engineering fees + licences	14·8
	Taxes	10·7
	Contingency	29·6
	Escalation	27·7
	Start-up cost	3·8
	Total plant	359·6
	Interest during construction	46·2
	Total	405·9[a]
Cost of service (1977)		
million $/year	Labour and supervision, tax, insurance, depreciation and return	81·2
	Repairs and maintenance	24·4
	Fixed charges	105·6[a]
	Electricity	9·9
	Boiler feed water	2·7
	Other utilities	1·2
	Catalysts and chemicals	5·7
	By-products credits	(9·7)
	Variable charges	9·8
	Total: million $/sd	0·350
	¢/MMBtu	139·9

[a] High inflation rates since the 1973 estimate now (mid-1975) cause this figure to be on the low side.

The maximum gasification efficiency in a Lurgi gasifier could be of the order of 70 % (though in practice it is less than this). Assuming a coal price of $10 per ton (US, 2 000 lb)* and a calorific value of 11 000 Btu/lb (*i.e.* a thermal cost price of 45·5 ¢/MMBtu) this corresponds to a gas send-out price of 140·0 + 65·0 = 205 ¢/MMBtu

* By early 1975, quoted US coal prices had risen to 40–50 $/ton.

(*see* also Fig. 11.2) on the assumption, of course, that the coal-to-SNG plant is operating continuously. (This estimate of gas price is to be compared with that of 360 ¢/MMBtu calculated for low temperature reforming of naphtha costing about 285 ¢/MMBtu—*see* Fig. 11.1.)

The relatively high investment and gas service costs of Lurgi SNG plants, coupled with their low gasification efficiency as compared with oil gasification plants has led to the development of a number of other

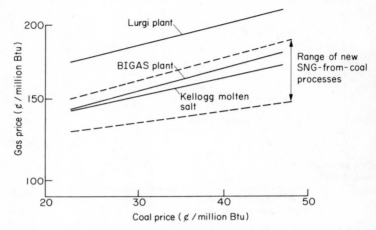

FIG. 11.2.   SNG cost as a function of process route and feedstock price—solid fuels.

more modern processes and orders have recently been placed for some of them. The following economic assessment[6] (see Table 11.7) refers to the BI-GAS process which was described in Chapter 9. Published information referring to a plant that would be operational in 1975 has been adjusted to take account of plant completion in 1977. Daily output of 250 million scfd requires three gasifiers in parallel.

The figures are indicative of slightly lower investment and reduced service charges as compared with the Lurgi process, though of course some reservations have to be made about the absolute values derived in each case.

Gasification efficiencies of the order of 75 %* have been estimated for full-scale BI-GAS SNG plants, and, again on the assumption of a coal supply at $10 per ton, the price of SNG produced in such a plant

* These are optimal estimates. In practice, initially at least, thermal efficiencies of these processes will be lower, and the gas price consequently higher than estimated here.

TABLE 11.7
Evaluation of the BI-GAS SNG Process

Investment (1976)
million $

Coal preparation and feeding	50·5
Shift conversion	24·9
Gasification	14·8
Acid gas removal and sulphur plant	23·4
Methanation and drying	23·6
Oxygen plant	52·6
Total on sites	189·8
Off sites and utilities	57·0
Total	246·8
Contractor's fee	12·4
Interest during construction, 10%	26·0
Contingencies, start-up	29·0
Total	314·2

Cost of service
million $/year

Capital charges; labour, supervision, tax, insurance	62·9
Operating charges; depreciation, return, repairs and maintenance	18·9
Fixed charges	81·8
Electricity	4·5
Boiler feed water	2·8
Other utilities	1·2
Catalysts and chemicals	5·7
By-product credit	(2·6)
Variable charges	11·6
Total, million $/sd	0·283
¢/MMBtu	113·2

is of the order of $113 + 61 = 174$ ¢/MMBtu. (The lower the conversion efficiency the higher the gas price.)

A third example of a coal-to-SNG process that has recently been costed is the Kellogg Molten Salt Coal Gasifier.[4] This is a route to SNG that has not so far been tested in a large pilot plant. However, it is fairly ingenious and differs sufficiently from most other coal gasification processes to make it desirable to evaluate its cost. The

TABLE 11.8
Estimate of Cost of Service of Kellogg Molten Salt Process

Investment		
million $		
	Coal preparation, transport and feeding	49·02
	Oxygen generation	52·40
	Shift conversion	24·93
	Acid gas removal and sulphur plant	23·42
	Methanation and drying	23·62
	Gasifiers	15·03
	Total on sites	188·42
	Off sites and utilities	56·53
	Total	244·95
	Contractor's fee	12·25
	Interest during construction	24·50
	Contingencies and start-up	28·50
	Total	310·20
Cost of service		
million $/year		
	Labour, supervision, tax, insurance, depreciation and return	62·04
	Repairs and maintenance	18·61
	Fixed charges	80·65
	Electricity	10·52
	Boiler feed water	2·80
	Other utilities	1·10
	Catalysts and chemicals	4·83
	Sulphur credit	(0·50)
	Variable charges	18·75
	Total, million $/sd	0·301
	¢/million Btu	120·5

version under consideration here is the single vessel gasifier in which coal is reacted with preheated steam and oxygen in a molten sodium carbonate bath. It will be remembered that, in a different version, air rather than oxygen can be used while hot melt is circulated between two vessels. However, the single vessel version appears to be further developed and therefore easier to assess economically.

Table 11.8 presents a summary of the steps in the investment and

TABLE 11.9
Comparison of Costs (Exclusive of Feedstock) of Coal and Petroleum SNG Processes

	Investment million $	Cost of service[a] ¢/million Btu
Coal		
Lurgi	405·9	139·9
BI-GAS	314·2	113·2
Kellogg molten salt	310·2	120·5
Naphtha		
CRG + hydrogasification	79·0	36·3
Gasynthan with two-stage	85·0	37·6
Gasification	80·0	37·0
GRH (+ LTR etc.)	108·7	46·7
Kerosine/gas oil		
GRH	115·0	52·6
Crude oil		
Partial oxidation + hydrocracking	250·0	94·4
Partial oxidation	171·6	71·2
FLEXICOKER + hydrocracking	247·3	88·7
Energy refineries distillate MDS + LSFO	180·0	72·4
Vacuum residue FLEXICOKING + distillate MDS	177·5	67·7
Vacuum residue partial oxidation + distillate MDS	189·0	87·0

[a] Exclusive of feedstock cost.

cost of service calculations, the estimates that are derived being not markedly different from those noted for the BI-GAS process.

Again one can calculate the cost of send-out gas for a US gas plant operating on open-cast coal of 11 000 Btu/lb heating value and costing say $10 per ton. Gasification efficiency will be about 77% and the SNG price is now 120·5 + 46·5 = 167 ¢/MMBtu.

The estimates of coal processing costs exclusive of feedstock charges are compared with those that were derived for petroleum-based processes in the summarising Table 11.9.

These show the oil processes to be significantly cheaper than the coal options in terms of both investment and service costs. However, as has already been indicated, the introduction of feedstock cost changes the picture on account of the markedly different price ranges applying to crude oil distillates and coal respectively at the time of writing. This topic is now discussed in greater detail.

11.6   EFFECT OF FEEDSTOCK PRICE

The cost of manufacturing SNG, it has been noted, is made up of the two elements of feedstock cost and cost of service. While cost of service can be predicted, albeit with some degree of uncertainty, over the next few years anyway, this is regrettably no longer true for many feedstock prices. Particularly crude oil and light distillate feedstocks have undergone such rapid and enormous price changes over the last three years that it would be next to impossible to forecast what they are likely to be over the next few years.

Similarly the price of coal not only has undergone a number of changes but varies widely between different areas. Future prices of mined coal are overshadowed by the rapid increase in mining wages, and price forecasts would, therefore, again be little more than conjecture. Consequently it is preferable not to regard feedstock price as constant but to plot as in Fig. 11.1 (liquid hydrocarbons) and 11.2 (coal) the total cost of SNG produced in different conversion processes as a function of a variable feedstock cost. In these figures, gas price (P) is related to feedstock cost (C), service cost (S) and gasification efficiency ($E\%$) by means of the formula

$$P = \frac{100\,C}{E} + S$$

where P, C and S are expressed in US cents per million Btu. The gas prices in the Figures (about 350 ¢ per million Btu, oil SNG v. 200 approx., coal SNG) correspond to maximum utilisation of the plant (330 days per year) and manufacturing costs for lower rates of utilisation, which will be higher, can be derived from the cost-of-service calculations in Tables 11.2 to 11.8.

However, it will have been noted that raw material cost (200–380 ¢/MMBtu), in the case of light distillate and crude oil conversion processes, is much higher than the cost-of-service (14–87) and that plant load factor is, therefore, less significant. For coal conversion, on the other hand the cost of service (120–140) is greater than that of feed (45·5) and a lower load factor would lead to a substantial increase in gas price, thus underlining the need for continuous operation of these plants. Absolute costs for coal-derived SNG at present day US coal prices are also lower than those for naphtha or crude-oil-based gases. Maintenance of this cost advantage in the future depends on the continuation of a cheap coal situation.

On grounds of both feedstock availability and SNG cost it would thus appear that coal-based SNG plants would be the ones most likely to be built in the US in the distant future. Against this long-term preference for expensive coal conversion equipment, the present-day selection of much cheaper naphtha- and crude oil-based SNG plants by many US utilities must be considered a convenient solution of the imminent gas shortage, which requires the immediate use of existing technology, at a time when gas selling prices are being forced upwards by shrinking natural gas reserves to levels matching those of our base load SNG estimates.

Only if costs of coal winning and transport become very high or if coal were to be unacceptable for environmental reasons are petroleum feedstocks likely to be considered as a long-term solution. Under those circumstances the limited availability of naphtha and middle distillates would tend to encourage the use of crude oil; however the differences in service cost between naphtha and gas oil has already been shown to justify a price differential of almost $1·0 per barrel. Similarly a light crude would have to be $3·0 per barrel cheaper than naphtha to produce SNG at the same price as the latter.

What is certain, however, is that oil processing can today provide an economic means of peak shaving natural gas, or of backing up incremental loads when these can only be offered on an interruptible basis because of diminishing natural gas supplies.

REFERENCES

1. Tennyson, R. N. and Buckingham, P. (1973). 'SNG Refinery' costs predicted, *Oil Gas J.*, **71**(26), 123.
2. Bodle, W. W. (1973). *Long Range Economics of SNG Production*, Paper 26, Inst. Gas Tech., SNG Symposium I.
3. Conser, R. E. (1973). *The Environmental Fuels Processing Facility*, Paper 16, Inst. Gas Tech., SNG Symposium, I.
4. Cover, A. E., Schreiner, W. C. and Skaperdas, G. T. (1973). Kellogg's coal gasification process, *Chem. Eng. Prog.*, **69**(3), 31–36.
5. Hazelton, J. M. and Tennyson, R. N. (1973). SNG refinery configurations, *Chem. Eng. Prog.*, **69**(7), 97–101.
6. Hegarty, W. P. and Moody, B. E. (1973). Evaluating the BI-GAS SNG process, *Chem. Eng. Prog.*, **69**(3), 37–42.
7. Lom, W. L. and Agius, P. J. (1975). *Technology and Economics of Clean Fuel Gas Manufacture from Liquid Petroleum*, 9th World Petroleum Congress, Tokyo, P.D. 17, Paper 1.

8. McMahon, J. F. (1973). *The FBH Process—SNG Production from Crude Oil*, Paper 13, Inst. Gas Tech., SNG Symposium I.

9. Morikawa, K., Najima, S. and Okagami, A. (1975). *Technology and Economics of SNG Manufacture from Naphtha*, 9th World Petroleum Congress, Tokyo, P.D. 17, Paper 2.

10. Wett, T. (1973). SNG from coal involves big projects, *Oil Gas J.*, **71**(26), 131.

11. Williams, F. G. and Schuller, R. P. (1973). SNG, naphtha and low sulphur fuel oils from crude using accepted technology, *Proc. Tech. Int.*, **18**(6/7), 265/6.

12. Tennyson, R. N. and Buckingham, P. (1973). What will it cost to make SNG from light hydrocarbons? *Oil Gas J.*, **71**(16), 99.

Chapter 12

A Glance into the Future

12.1 INTRODUCTION

In this chapter, we try to envisage how SNG may fit into future world energy patterns, having regard to trends in production, consumption and new technology that can be discerned at the present time. This means starting of course with some of the uncertainties. Most importantly, in respect of our subject topic, SNG, part of the technological discussion in this book is built on laboratory or pilot plant studies that still had to be completed, namely the work on hydrogenation of heavy oils, gasification of coals and the 'energy refinery' concept. Only a relatively small proportion can be said to relate to published results of full scale demonstrations in commercial size equipment, for example gasification of refinery naphthas.

We have already begun the process of forecasting in previous chapters and our present task means going beyond the limits in some instances of immediately foreseeable developments.

Furthermore, technological change, particularly in the energy field, is almost inevitably the direct result of pricing policies in regard to raw materials or substitutes. And while price developments can normally be predicted, at least directionally in a free market economy on the basis of supply and demand, recent events have shown that the days of monopolies and price control by small groups of producers are by no means over. The petroleum price rise in 1973/74 by a factor of three and a half was totally unforeseen and any repetition upwards or downwards could make technical forecasting a somewhat unprofitable exercise.

It has therefore become fashionable to hedge one's economic and technological bets and to forecast in terms of different 'scenarios', the choice of high or low energy prices, large or small availability of fossil fuels, high or low cost of capital being left to the reader. The authors consider this approach not only totally useless but also rather dangerous. After all, what is there to stop a reader from selecting a particularly unlikely combination of circumstances and to arrive at an even unlikelier conclusion—and worse still, possibly to act on it.

We prefer to describe what we see as being the most likely events, for which purpose we must combine some positive statements of fact with some unambiguous assumptions.

12.1.1 Energy Growth
We can assume that despite present predictions of zero to slow growth economies, an overlay of boom times will result in overall growth in energy consumption per head. Thus will continuing pressures for improved living standards be met and thus, with an ever increasing population growth, will total energy demands continue to rise.

12.1.2 Fossil Fuels
The world will continue to rely mainly on fossil fuels, and reserves will be sufficient to meet demands for at least another 100 years. Alternative energy sources such as fuel cells, solar power and geothermal energy (*i.e.* via deposits of pressurised hot water), tidal and wind power* will have little or no impact inside the next 25 years.[21]

12.1.3 Petroleum
Crude oil and derived products will continue to be relatively expensive owing to producer monopoly action and because of the need to tap high-cost fields in not readily accessible locations. Therefore, in future, usage of petroleum products will become concentrated in those outlets that require—and can afford—its special properties, for example, transportation fuels, petrochemicals, premium industrial applications, food production, etc.

12.1.4 Natural Gas
Declining rates of wellhead production coupled with conservation measures will result in natural gas supplies from the traditional sources and relatively new finds being insufficient to meet demands for clean-burning gaseous fuel. What little incremental gas is found will be in remote locations at great depths and often offshore. Shortages coupled with higher production costs will force prices upwards.

---

* Environmental effects of wind turbine systems are being studied by Battelle Laboratories, Ohio, using a 100 kW generator with a 125-ft diameter rotor blade mounted on a 125-ft tall tower (*Energy World*, Oct. 1975, No. 20, p. 11).

12.1.5 Coal

Supplies of coal will no doubt outlast those of oil and though prices will be appreciably higher than today's, they should not reach those pertaining to either oil or gas. Coal therefore will be converted into liquid and gaseous fuels. It will continue to supply present needs for electricity but probably few, if any, new coal burning power stations will be built after the end of this century.

12.1.6 SNG

In the immediate future, natural gas prices will remain low enough to prevent sizeable quantities of SNG being manufactured by the available oil processing technology except in areas critically affected by shortages and perhaps to back up interruptible natural gas supply contracts. As coal processing technology improves and cheapens the necessary investment, coal will become the preferred feedstock. Full exploitation, however, will be prevented by shortages in the coal winning and handling sectors.

Therefore, means of reducing SNG costs and additional ways of bridging the gas demand/supply gap will be sought. Provided present concern over safety aspects of nuclear power generation are overcome—and this seems inevitable—atomic fission, and much later fusion—will open up new possibilities. Thus a combination of electricity and heat raised from nuclear sources would speed up the generation of hydrogen for utility and industrial purposes from coal and water respectively. Also nuclear heat could be made available to supply the endothermic heat necessary for the gasification of coal or crude oil.

However, before the advent of the so-called hydrogen economy, we see the clean gas gap being closed in a number of other ways. Firstly, more emphasis on supplying factory heat from higher sulphur fuels, fuel oil and coal, using means of removing the sulphur from combustion gases or stack effluent so as to prevent environmental pollution; secondly, a return to lean gas for pipeline supplies to captive industrial customers; thirdly, use of methanol, generated from fossil fuel resources far from centres of consumption, and transported to those same centres such as LNG is today, as a feedstock for SNG manufacture as well as for automotive and its present industrial chemical purposes.

The topics of SNG production from methanol and from coal with the assistance of nuclear heat, the creation of industrial lean gas grids and hydrogen distribution are selected for elaboration in the closing sections of this book.

12.2   LEAN GAS FOR USE IN INDUSTRY

While the conversion of solid and liquid hydrocarbons into lean gas, defined as having a calorific value of 100 to 450 Btu/scf (955–4300 kcal/m^3), does not strictly belong in a text on SNG, the calorific value of which is at least 800 Btu/scf (7640 kcal/m^3), there are certain aspects of lean gas manufacture that merit some passing comment. In particular, both SNG manufacture and lean gas manufacture result in clean-burning, sulphur-free fuels and they could therefore be used interchangeably for many, though not all, applications, for example industrial processing, which frequently uses a limited number of very large burners, not always tied to a particular gas composition. Furthermore, burners can be modified and adjusted and, provided gas quality remains constant between changeovers, there is no reason why one particular type of gas should be used on an exclusive basis.

Lean gases are obtained from liquid and solid hydrocarbons by one of two processes that have been discussed in greater detail in previous chapters. Firstly, light hydrocarbons such as natural gas, refinery gases, LNG and particularly naphtha and gas condensate are steam reformed at temperatures of about 800 °C in the presence of a catalyst and undergo a reaction that can be summarised as:

$$C_6H_{14} + 6H_2O \rightarrow 6CO + 13H_2$$

The resultant gas has a calorific value approximating to the mean value for carbon monoxide and hydrogen (320 Btu/ft^3–30 in Hg, 15 °C sat'd; 3100 kcal/Nm^3).

Secondly, almost all hydrocarbons, including fuel oil, crude oil, and coal, irrespective of molecular weight, can be reacted with oxygen and steam or air and steam, at temperatures in the range 1100–1400 °C to produce, again, a mixture of hydrogen and carbon monoxide, with some carbon dioxide present, and, of course, diluted with nitrogen if air is the oxidant.[2] In this partial oxidation process, the calorific value would again be about 320 Btu/ft^3 if oxygen were the oxidant but with air the value would be reduced to about 125 Btu/ft^3.

As far as industrial customers are concerned there are three aspects of gasification that are of prime interest. Of these the absence of polluting waste gases has been mentioned; the other two are efficiency of conversion of the solid or liquid fuel into a gas and of transmission of the gas to the customer; and capital cost. While low Btu gases match SNG in respect of pollution there are apparent differences as far as gasification/transmission efficiency and gasification investment are concerned.

Gasification efficiency, usually defined as the ratio of the calorific value of the manufactured gas to that of the total feed (the latter also including process fuel, external steam and oxidant), varies widely from process to process and even between plants using the same process. It is meaningless to compare oxygen-consuming processes with others that operate on air; a high power consumption can give a false impression of higher efficiency; by-products may or may not be usable as boiler fuel; chemicals or solvents and their yields may vary. Sulphur in the feed contributes to its calorific value but is removed in the course of gasification. Finally, the degree of heat integration of the gasifier, and thereby its cost, vary from plant to plant. It is, therefore, preferable in our view to compare theoretical efficiencies of different processes, rather than figures published in the literature, which frequently are not too well defined.

As soon as we confine ourselves to theoretical considerations, however, we note that the overall efficiency of a complete conversion from hydrocarbon to the combustion products CO_2, H_2O and nitrogen should be the same, irrespective of the intermediate gaseous fuels, mainly carbon monoxide and hydrogen in the low Btu case, mainly methane in SNG manufacture. Differences between processes will be entirely due to the occurrence of heat losses and the formation of non-gaseous by-products.

In both these respects, low Btu gases seem to score over SNG. The greater complexity of SNG plants often results in higher losses, and the synthesis of methane frequently results in the formation of by-products such as aromatics and char. On the other hand the higher temperatures of the low Btu processes, unless complex heat exchange facilities for furnace and product gases are provided, will result in reduced efficiency, and the formation of char by thermal cracking must be prevented by careful preheater design to avoid additional process heat losses. As a result, high temperature reformers and partial oxidation plants, although inherently less complex than SNG facilities, require investment of the same order, particularly if the

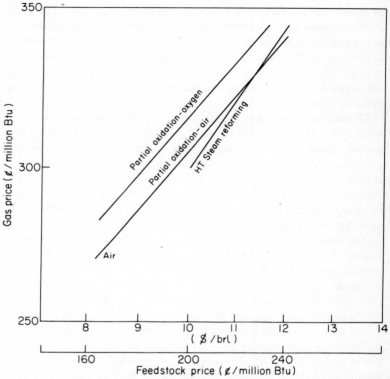

FIG. 12.1   Lean gas cost as a function of process route and feedstock price. (Source: Ref. 14.)

latter is expressed as investment per unit calorific value. In fact both the gasification losses and the capital cost of low Btu gas production are only marginally lower than those of SNG facilities.

On the other hand the operating cost of handling and distribution of low Btu gas is undeniably higher. Piping, compression and purification costs are all functions of the volumes of gas handled and the higher calorific value of SNG conveys a distinct advantage on the latter. Even throughput capacity of a system, which depends on Wobbe Index rather than calorific value, is much higher for SNG than for some low Btu gases, though not so in comparison with hydrogen-rich gases.

The cost of producing low Btu gas from petroleum feedstocks in plants of the same thermal capacity as those mentioned in Chapter 11 is summarised in Fig. 12.1 and, it will be noted, is little different from that of SNG (*see* Fig. 11.1, 285–340 c/MMBtu for feeds costing 8–12 \$/brl). Partial oxidation with either oxygen or air is preferable if fuel prices are high; at lower feedstock prices steam reformed gas tends to be cheaper.[14]

In conclusion, it would seem that an incentive to use low Btu gas rather than SNG exists only where the gas in question is used close to its point of generation, where its manufacture is efficiently integrated and heat losses are minimised, and where its non-polluting characteristics are at a premium.[16] The laying of low-Btu gas mains in concentrated highly industrialised areas can, therefore, be visualised as an interim measure under the circumstances discussed in the introduction to this chapter.

12.3 METHANOL AS A SOURCE OF ENERGY AND AS FEEDSTOCK FOR SNG

The cheapest and most obvious range of raw materials for the manufacture of SNG are clearly the naturally occurring hydro-carbons. Where these have to be processed or treated before gasification it is desirable that such operations should be simple and cheap—the fractionation of crude oil and the production of a naphtha gasification feedstock is a case in point—rather than complex and relatively expensive, as is the case with conversions into chemical intermediates.

While this rule applies fairly generally, there exist already, and will arise more frequently in future, circumstances that might justify the conversion of a natural raw material into a chemical compound that would then be used to manufacture SNG. For such a process route to be technically and economically attractive a number of preconditions should be met:

—the manufacture of SNG should take place in equipment that is equally suitable for other feedstocks such as naphtha,
—the intermediate chemical should be cheap, *i.e.* its manufacture should be easy and the energy conversions involved should be thermally efficient,
—the intermediate chemical should be easily transported and stored.

One of the few chemical commodities that meet most of the above stipulations is methyl alcohol (methanol), and considerable thought has been given in the recent past to large-scale methanol manufacture from either surplus natural gas or coal and subsequent gasification of the alcohol to produce a natural gas substitute.[1,5,13]

The manufacture of methanol from natural gas is simple in theory but fairly complex in practice. Natural gas is steam reformed under pressure at a temperature of 800–820 °C to produce a mixture of carbon monoxide and hydrogen. Excess hydrogen is reduced or, put differently, the proportions of hydrogen and carbon monoxide in the gas are adjusted to a ratio of about 2:1, and the gases are reacted over a catalyst, to form methanol, in accordance with the series of reactions below. While methanol synthesis at one time required a minimum pressure of 150 atm, the invention of modern catalysts has resulted in a gradual reduction of operating pressures; recently built plants synthesise methanol at 50 to 60 atm, and can be designed without a synthesis gas compressor by operating both steam reformer and CO conversion at a pressure of about 60 atm.

$$CH_4 + H_2O \rightleftharpoons CO + 3H_2$$
Reforming

$$CO + H_2O \rightleftharpoons CO_2 + H_2$$
CO conversion

$$CO_2 + CO + 3H_2 \rightleftharpoons 2CO + 2H_2 + H_2O$$
Hydrogen reduction

$$CO + 2H_2 \rightleftharpoons HCH_2OH$$
Methanol synthesis

The regasification of methanol to SNG, on the other hand, does not require equipment substantially different from that used for naphtha gasification.[2,7,8] Low temperature steam reformers such as the CRG, MRG or Gasynthan process plants discussed in Chapter 6 will produce SNG from a methanol feed by a complex series of reactions that can be expressed very simply as:

$$4CH_3OH \rightarrow 3CH_4 + 2H_2O + CO_2$$

This reaction, however, is considerably more exothermic than low temperature reforming of naphtha feeds and precautions must be taken to prevent carbon formation by cracking or by the Boudouard reaction ($2CO \rightleftharpoons CO_2 + C$). There are three possible means of keeping the reaction under control:

—multistage gasification with intermediate cooling of the gasification products,
—external cooling of the gasification reactors,
—recycle of cooled reaction products, either with or without prior condensation of steam.

Since the first two approaches necessitate process equipment somewhat different from that normally available in naphtha gasifiers, most attention has been paid to the third approach, particularly the recycling of wet product gas, which supplies the required steam dilution and thus obviates the need for additional process steam.

A final product gas, after single-stage methanation and CO_2 removal made up of:

methane	97·80%
carbon monoxide	0·05%
carbon dioxide	0·10%
hydrogen	2·05%

and having a gross calorific value of 9420 kcal/Nm^3 (997 Btu/scf) has already been produced in experimental equipment.[5]

Clearly, in order to produce methanol cheaply its raw material, natural gas or coal, must be available in large quantities and must also be very cheap. Furthermore, conversion costs must be kept to a minimum, which implies that methanol must be produced in a very large, modern and efficient plant.[17] Finally, transport cost of the chemical from manufacturer to gas distributor must be kept low, and this can only be achieved by sea transport in very large ships.[19]

It follows that methanol manufacture from natural gas, to have any chance of economic success, will have to be carried out in overseas countries with a large production of associated natural gas, local outlets for which are limited by lack of industrialisation and production of which is inevitable as it has to be produced alongside the crude petroleum which is the primary product.

It should be borne in mind, however, that the chain: natural gas–methanol manufacture–methanol transport–gasification–SNG starts with the same raw material and finishes with the same end product as does the liquefied natural gas scheme: natural gas–liquefaction–LNG transport–regasification–natural gas. A decision whether the first or the second route to methane should be used will, therefore, depend on the relative economics of the two cycles. Most techno-economic studies of the two alternatives have led to the conclusion that the larger the shipping distance between gas supply and distribution areas, the more attractive methanol becomes as an intermediate. Critical distances below which LNG and above which methanol becomes more attractive have been variously estimated as 4000, 5000 and even 7000 miles, leaving only the longest LNG runs as possible candidates for methanol conversion. No large

TABLE 12.1
Economics of Methanol versus LNG Shipment
Movement of $5·0 \times 10^6$ tons of Methane Equivalent

	LNG schemes		Methanol schemes	
	12 000 miles	6 000 miles	12 000 miles	6 000 miles
Investment, $10^6\$$				
Process plant	160	160	321	321
Off-site facilities	81	81	83	83
Start-up, spares, royalties	20	20	42	42
Plant storage, mooring, loading	70	70	15	15
Working capital	33	33	42	42
Ships	1 100[a]	640[b]	275[c]	140[d]
Terminal storage, unloading	70	70	15	15
Regasification	5	5	25	25
Total	1 539	1 079	818	683
Total ex ships	439	439	543	543
Gas price, $¢/10^6$, Btu				
Wellhead price	10·0	10·0	10·0	10·0
Process fuel and loss	0·3	0·3	8·2	8·2
Evaporation loss	3·0	3·0	—	—
Capital cost: ships (13%)	53·6	26·8	13·4	6·8
all other (20%)	32·9	32·9	40·7	40·7
Operating cost: manufacture	4·8	4·8	11·0	11·0
shipping	15·0	7·5	10·0	5·0
other	2·0	2·0	8·0	8·0
Total	121·6	87·3	101·3	89·7

[a] Ten LNG carriers, 200 000 m³, average speed 16 knots.
[b] Five LNG carriers, 200 000 m³, average speed 18 knots.
[c] Eleven VLCCs, 200 000 tons, average speed 16 knots.
[d] Six VLCCs, 190 000 tons, average speed 16 knots.
All investments based on Ref. 19.
Source: Ref. 13.

methanol plant has, in fact, been built so far in any of the gas surplus countries; but projects have been evaluated for a number of locations including Iran, Saudi Arabia, Abu Dhabi, Kuwait, Libya and Algeria.

One of the reasons for the relative lack of success of methanol schemes as compared with LNG schemes, in the recent past, is the oil price increase and the much higher notional cost of natural gas feedstocks. Since the lower transport cost of methanol, as shown in Table 12.1, is balanced by the higher process fuel and loss item, an increase in the relatively low wellhead price of gas assumed in this 1973 study immediately results in a shift of the economic equilibrium away from methanol and towards LNG.

The widespread adoption of methanol as a means of transporting natural gas across the oceans and, therefore, the availability of methanol as an SNG feedstock has now probably receded further into the future than it was a few years ago. Nevertheless it is possible that improvements in process technology that would reduce methanol manufacturing cost, on the one hand, and even longer shipping distances and the high and still rising cost of LNG ships and LNG transport, on the other, might shift the economic balance between the two the other way in the future.

The manufacture of methanol from coal is of necessity more complex than from natural gas. Nevertheless it is conceivable that, under the impact of limited supplies of high priced petroleum and LNG, coal could become of interest as a raw material for methanol production. However, the economics of coal-based methanol as a feedstock for SNG would depend critically on overall conversion cost and efficiency of the coal to SNG via methanol route being more favourable than direct coal gasification. This will only be the case if

—coal is extremely cheap, oil and gas are in short supply,
—gas transport is very expensive,
—coal gasification processes are less efficient than methanol manufacture and gasification,
—SNG is required for peak demand only and storage is therefore at a premium,
—methanol is mainly required as a motor fuel for internal combustion engines and SNG feed is merely a by-product.

It will of course be rare for all these preconditions to apply at present and unless economic conditions change drastically, we do not expect any large-scale methanol developments from coal in the immediate future. In the longer term, however, this may become an acceptable route to SNG.

12.4   NUCLEAR ENERGY ASSISTED GASIFICATION PROCESSES

The gasification of all hydrocarbons and particularly of fuel oil or coal by means of steam has been shown to be highly endothermic.[2] Similarly the production of hydrogen, which can subsequently be used in exothermic gas processes, requires itself an input of energy. In the past the energy deficiency of these gasification routes has always been met by consuming part of the hydrocarbon feedstock as fuel or by supplying the required energy separately in the form of other fossil fuels.

However, in the light of the predictions made in the introduction to this chapter, this may no longer be the cheapest route in future. Even at today's fossil fuel prices electric power generated in modern nuclear reactors is becoming noticeably cheaper, particularly as far as base demand is concerned, than that produced in oil fired plants. Similarly European coal may eventually price itself out of the energy market, and unless a substantial drop occurs in world fossil energy prices, thermal energy derived from nuclear fission—and possibly later from nuclear fusion—will soon become the cheapest form of heat available.[4,20]

This does not imply that SNG and other gases will no longer be required. Their ease of transportation, flexibility in use, rapid heating power, chemical reducing properties, and the fact that distribution networks for gases are available, should result in SNG production even at a point in time when nuclear electricity generation has cheapened even further and spread even wider than today.

However, under those circumstances substantial savings in cost could be achieved if the part of the fossil fuel consumed in a gasification process purely in order to supply the endothermic reaction heat could be replaced by a nuclear heat source. Since both hydrogen manufacture and the steam reforming reactions used in gasification processes are endothermic two basic process routes to nuclear energy assisted gasification can be envisaged. Development work on both nuclear hydrogasification and on the gasification of lignites and bituminous coals with steam is in fact in progress at the moment, mainly in Germany.[12]

In order to achieve reasonable reaction rates the endothermic reaction between a hydrocarbon fuel and steam is normally conducted in the temperature range of 700–800 °C. To assist the process by means of a nuclear energy source it is, therefore, essential to make heat available at a somewhat higher temperature, about 900 to 950 °C, in other words to employ a high temperature nuclear reactor.

It is not proposed to discuss the principles of high temperature gas-cooled nuclear reactors in any detail in this context. Suffice it to say that these reactors burn enriched uranium, that the cooling gas is helium and that several experimental reactors at Peach Bottom, USA, Winfrith, UK, and AVR, Germany, and a power plant at Fort St Vrain in the USA, have been operating for several years. Gas exit temperatures at most of these plants would, however, be insufficient for high temperature processing. Only the modified experimental reactor at Jülich, Germany, so far, which has been in operation for about one year, produces helium at 950 °C, the temperature envisaged for SNG plant heating. A 300 MW plant for coal gasification based on 750 MW of thermal energy to be produced in a high temperature gas-cooled reactor is now being designed by Arbeitsgemeinschaft Nukleare Prozesswärme (ANP), and points the way ahead for using nuclear heat in commercial fuel gasification projects.

A number of operating schemes for nuclear assisted coal gasifiers have consequently been proposed by ANP[4] and two examples of a lignite hydrogenation and a steam gasification process for bituminous coal respectively can be quoted.

Lignite is dried and contacted with hydrogen-rich recycle gas in a gasifier at about 900 °C. The feed is thereby devolatilised and hydrocracked with some exothermic methane formation. Any residual char formed in the gasifier is withdrawn and discarded. The gases are purified by removal of H_2S, CO_2 and organic sulphur compounds. This leaves mainly hydrogen, carbon monoxide and methane, which are separated by gas fractionation. The methane–SNG product is drawn off while the hydrogen/carbon monoxide mixture is recycled through the helium heat exchanger, where its temperature is raised to 900 °C, back to the gasifier.

Hot helium from the nuclear reactor is cooled first from 950 to 900 °C in the gas-to-gas heat-exchanger then in a steam generator from 900 to 250 °C. Part of the steam produced is added to the recycle gas upstream of the heat exchanger; the remainder goes to a steam turbine power generator and is condensed and recycled.

The second scheme envisages the complete gasification of bituminous coal with the aid of steam raised to a temperature of 900 °C, by means of heat exchange with helium that leaves the nuclear reactor at 950 °C. Bituminous coal is pulverised and introduced through a lock into a pressurised low temperature carbonisation vessel. The coal is devolatilised at 800 °C by contact with a series of coils carrying helium gas, and passes through a cracker into a fluid-bed gasifier where it reacts with superheated steam. The gasifier is maintained at 850 °C, again by circulating secondary helium through coils immersed in the fluid bed. Coal volatiles, partly cracked in the previous vessel, and the gasification products of steam and reactive char are filtered, steam converted, purified by H_2S and CO_2 removal and methanated to produce SNG. Two helium circuits, a primary circuit at 950 °C, and a secondary circuit at 900 °C, heat the carbonisation vessel and gasifier and produce high temperature process steam and also generate superheated steam for a power turbine.

It will be appreciated that these are just two out of many technically, though not at present economically, feasible routes to SNG. There is little doubt, however, that economically favourable conditions will come about in the not very distant future under which some such scheme could be put into practice, if the events that we discussed in the introduction materialise.

12.5   NUCLEAR HEAT TRANSPORT

Although the chemical conversions that we propose to discuss under this heading are not strictly connected with SNG manufacture, the fact that the methanation of carbon monoxide to methane is involved establishes, in our view, a sufficient link. One of the difficulties of thermal energy generated in nuclear reactors is its lack of transportability. High temperature reactors are by definition large and bulky and yet they should, for safety reasons, be located some distance from other industrial plant and built-up areas. While heat energy transport in the form of steam, hot water, electricity or compressed gas is technically feasible, transport losses, particularly over long distances, are substantial and a better method of energy transfer would be desirable.

A suggested means[12] of transferring heat from a high temperature reactor to remote consumers is to convert methane by high temperature catalytic steam reforming into a mixture of hydrogen and carbon monoxide.

$$CH_4 + H_2O \rightarrow 3H_2 + CO$$

The resulting synthesis gas can then be piped at ambient temperature and with minimum loss over substantial distances. On reaching its destination it can either be stored, a particular advantage when compared with electricity which has to be used immediately, or it can be passed through a methanation reactor where it undergoes the reverse reaction

$$CO + 3H_2 \rightarrow CH_4 + H_2O$$

By selecting the right conditions of catalysts, temperature and pressure both reactions will go close to completion and endothermic heat absorbed from the atomic pile in the reforming reaction will be fully released in the methanator. Methane reformed in the process can either be dried and returned by separate pipeline to the HTR, since water need not be transported back, or be used locally as a high grade fuel. In either case, it is claimed that overall efficiency of the scheme is substantially higher than that of competing methods of transferring heat energy from a high temperature nuclear source to users some distance away.

Nevertheless some of the problems of such a scheme should be borne in mind. Temperature of the recovered heat is limited to around 300 °C if recombination is to be fairly complete. Catalyst deactivation would have to be avoided, particularly of the methanation catalyst on the energy consumer's premises, and the need for two pipelines, if methane is to be returned, would add to the cost of such a scheme. It should also be borne in mind that low grade energy is often available cheaply as a by-product of power generation and that outlets for the proposed piped energy supply may turn out to be few.

Nevertheless there may well arise instances where temperature, distance from the nuclear plant and, possibly, the need for some energy storage may make the piped transport of carbon monoxide and hydrogen to a remote energy consumer an attractive proposition, particularly when shortages and high prices of clean-burning fossil fuels will have improved the relative economics of nuclear heating.

12.6   HYDROGEN AS A SUBSTITUTE FOR NATURAL GAS

An obvious conclusion to be derived from our picture of the gradual disappearance of natural gas, liquid petroleum and coal from the energy scene is the need to replace natural gas ultimately with a non-fossil fuel. Once gas and oil are no longer available and coal has become too valuable for conversion into pipeline gas, hydrogen becomes practically the only gaseous fuel that can still be manufactured in large volumes from the remaining energy resources of the world.[10]

While the cheapest route to hydrogen at the moment is still its manufacture from solid, liquid or gaseous hydrocarbons by steam reforming or by partial oxidation with oxygen, both followed by shift conversion and CO_2 absorption, as discussed in Chapters 7 and 9, there are a number of other manufacturing processes that do not require the massive availability of hydrocarbons. Naturally, if we assume that hydrocarbons are either unavailable or too expensive it is these processes that will have to be employed.

The raw material for the alternative preparation of hydrogen in large volumes must evidently be water which, in order to convert it into hydrogen, must undergo a highly endothermic chemical reaction, the energy deficit of which can be supplied in various forms, generally in the form of electric power, but as more recently envisaged, and with a view towards improving the efficiency of the reaction, also as thermal energy.

The manufacture of hydrogen from water by means of electricity is, of course, well known. Electrolytic dissociation of water into hydrogen and oxygen has been practised for many years[11,15] and numerous designs of electrolytic cells, electrodes, electrolytes, gas collection devices, etc. have been commercialised. Temperature of the electrolyte affects power consumption, and high temperature hydrogen cells and also high pressure equipment have reached an advanced stage of development.[6,9] Electrolytic preparation of hydrogen should preferably be located close to a power plant, in order to avoid transmission losses, and outlets must also be found for the large volumes of oxygen produced simultaneously.

To avoid the inherent inefficiency of electric power generation some form of thermochemical dissociation of water has frequently been proposed as an alternative to electrolysis. Unfortunately the unaided dissociation of the water molecule requires a temperature of about

2000 °C and would therefore be both uneconomic and inconsistent with our picture. There are, however, a number of chemical reaction series, mostly based on Redox systems, that allow one to reduce this temperature to between 650 and 900 °C, temperatures that are attainable with the aid of nuclear reactors and therefore independent of the availability of fossil fuels. One such system named after its inventor the Marchetti Cycle (mark 9) is the following:

$$6FeCl_2 + 8H_2O \rightarrow 2Fe_3O_4 + 12HCl + 2H_2 \uparrow \quad (at\ 650\ °C)$$

$$2Fe_3O_4 + 3Cl_2 + 12HCl \rightarrow 6FeCl_3 + 6H_2O + O_2 \uparrow \quad (150–200\ °C)$$

$$6FeCl_3 \rightarrow 6FeCl_2 + 3Cl_2 \qquad and\ so\ on....$$

The thermal dissociation of water, based on this cycle, would thus consist of a process stage at 650 °C during which ferrous chloride would produce hydrogen and hydrochloric acid by reacting with steam, followed by a second stage in the course of which condensed hydrochloric acid and ferric oxide combined at 150 to 200 °C to reform ferrous chloride. A number of other 'carriers', in addition to ferrous chloride, have been proposed and, theoretically anyway, there seems no reasons why, even now, cheap thermal energy should not be utilised to produce bulk hydrogen by means of such a cycle. At a later stage, when the supply of fossil fuels had become more restricted, hydrogen from this source would have to take on the task of replacing natural gas or other piped gas supplies around the world.

The question immediately arises: how does hydrogen compare with existing piped natural gas (methane) in terms of physical properties and combustion parameters? Reference to Table 12.2 shows that there are wide differences, some of which work in favour of hydrogen, others against it.

The prime advantage claimed for hydrogen is an environmental or ecological one. Hydrocarbons produce carbon dioxide, water vapour and nitrogen on combustion; hydrogen on the other hand yields water vapour and nitrogen only and the former rapidly condenses to refurbish the earth's water reserves, which in turn provide the source for further hydrogen production by electrolysis or other forms of dissociation.

In addition, hydrogen burns cleanly (Delbourg yellow tipping index 0 v. 134) and very efficiently because of its low stoicheiometric air demand ($2·4\ m^3/m^3$ v.$9·5$) and decreased stack gas volume ($2·9\ m^3/m^3$ v. $10·5$). Its high burning velocity and flame temperature permit the

TABLE 12.2
Hydrogen, Possible Alternative Fuel to Natural Gas
Comparative Data

	Hydrogen	Natural gas (methane)
Molecular formula	H_2	CH_4
Boiling point, °C	−252·8	−161·5
Liquid density, g/l (at °C)	71 (−252·8)	425 (−161·5)
Relative density, 15 °C, 1 atm (air = 1)	0·069 6	0·542
Calorific value (gross)		
kcal/g mole, 25 °C	70·4	216
kcal/kg	35 200	13 500
Btu/ft^3	325	1 000
kcal/Nm^3	3 050	9 572
Products of combustion, m^3 per m^3 of gas		
CO_2	Nil	1·0
H_2O	1·0	2·0
N_2	1·9	7·5
Ultimate CO_2 in flue gas (%)	Nil	11·8
Wobbe Index, metric	11 714	12 997
Flame speed factor (Weaver)	100	14
Yellow tipping index (Delbourg)	0	134
Ignition temperature, min °C in air	570	470
Inflammability range, gas/air, % vol		
lower limit	4	5
upper limit	75	15
Flame temperature, °C in air	2 200	1 920
Specific heat at STP, cal/mol/°C	6·9	8·2
Liquid/gas vol/vol expansion ratio	865	650

use of small combustion chambers and high intensity flames. Its low specific gravity makes for rapid and easy dispersal of escaping gas.

On the debit side, in comparison with natural gas, hydrogen has a low calorific value (3050 kcal/m^3 v. 9572) and if stored occupies a larger volume (for equal thermal potential); a higher moisture production rate for equal heat release (0·35 m^3 v. 0·22) so that, used as an interior space heating fuel it produces more condensation. Its inflammability limits are wider, i.e. ignition is easier because, although its ignition temperature is higher, ignition energy requirement is only one tenth that of methane. Its Wobbe Index is slightly lower, implying a small reduction in the carrying capacity of pipelines and distribution grids.

Storage of both natural gas and hydrogen present problems since compression is expensive and liquefaction is only possible at extremely low temperatures, the atmospheric boiling point of hydrogen ($-252 \cdot 8\,^{\circ}$C) being even lower than that of methane ($-161 \cdot 5\,^{\circ}$C). However, apart from liquefaction of the gas, which results in a density increase of 1:836, hydrogen, as distinct from methane, can also be stored in the form of metal hydrides. Certain metals, such as lanthanum–nickel alloys, will absorb selectively up to 5% wt of hydrogen at ambient temperature and low pressures, and will release the gas when heated over certain temperature ranges. While metal hydride storage at present requires expensive and rare metals there is work in progress to substitute cheaper and more widely available metals for the rare alloys used hitherto.[3]

Hydrogen when compared with methane can thus be shown to perform equally well as a fuel gas in most respects. It has in addition certain uses as a chemical, as a reducing agent, and a fuel for power generation in fuel cells which methane has not and which might speed up its adoption as a replacement for SNG, even before fossil fuels run out or become prohibitively expensive.

12.7   CONCLUSIONS

Our glance into the future has been rewarding in some respects, disappointing in others. More immediate developments such as the use of methanol or of lean gases for industrial combustion can be assessed and analysed with reasonable accuracy, but no such claim can be made in regard to longer term developments, many of which are based on the availability of cheap and abundant nuclear energy. The main weakness of any analysis of these more distant and less predictable events is the need to view them in a particular setting involving forecasts of both availability and price of the different fossil fuels at certain times. It is quite unavoidable that the accuracy of our technological crystal ball gazing effort should depend critically on the correctness or otherwise of our world energy picture. How wrong we were only time will tell.

REFERENCES

1. Anon. (1973). Methanol *v.* LNG, *Petroleum Press Service*, Feb.; Anon. (1973). Humphreys & Glasgow Look at Methanol Route, *Process Engineering*, 9.

2. B.P. Trading Ltd (1972). *Gas Making and Natural Gas*, B.P. Trading Ltd, London.
3. Battelle (1973). *Hydrides: A Practical Means of Packaging Hydrogen*, Information Notice BI 73.5, Geneva Research Centre.
4. Bergbau-Forschung GmbH, Essen, Rheinishe Braunkohlenwerke AG, Köln, Kernforschungsanlage Jülich (1973). *Sicherung der Energie und Rohstoffversogung durch nukleare Prozesswärme, Essen, Köln, Jülich*.
5. Blythe, B. M. and Hartley, W. (1973). *Methanol for the International Transportation of Energy*, Paper 11, Inst. Gas Tech., SNG Symposium I.
6. Borkris, J. O'M. (1972). *Electrochemistry of Cleaner Environments*, Plenum Publishing Corp., New York.
7. Cairns, J. and Hartill, I. J. (1973). *British Gas Experience of High Pressure Gas Making Processes*, 5th Synth. Pipeline Gas Symposium, Chicago.
8. Cairns, J. (1975). Future developments in SNG, *Gas Engineering and Management*, **14**, 361–373.
9. Costa, R. L. and Grimes, P. G. (1967). Electrolysis as a source of hydrogen and oxygen, *Chem. Eng. Prog. Symposium Ser.* 63, No. 17, p. 45.
10. Gregory, D. P. (1973). The hydrogen economy, *Scientific American*, **228**(1), 321–329.
11. Hampel, C. A. (1964). *The Encyclopaedia of Electrochemistry*, Reinhold, New York, pp. 1156–1160.
12. Harth, R. and Fehlhaber, K. (1973). *EVA—eine halbtechnische Versuchsanlage zur Untersuchung der Methanreformierung mit Wasserdampf unter Ausnutzung von Wärme aud einem Hochtemperaturreaktor*, Deutsches Atomforum, Bonn.
13. Lom, W. L. (1974). *Liquefied Natural Gas*, Applied Science Publishers Ltd, Barking.
14. Lom, W. L. and Agius, P. J. (1975). *Technology and Economics of Clean Fuel Manufacture from Liquid Petroleum*, 9th World Petroleum Congress, Tokyo, P.D. 17, Paper 1.
15. Mantell, C. C. (1960). *Electrochemical Engineering*, McGraw-Hill, New York.
16. Moss, G. and Agius, P. J. (1975). *The Desulphurisation of Fossil Fuels During Combustion in Fluidised Beds of Lime Particles*, 9th World Petroleum Congress, Tokyo, P.D. 18, Paper 3.
17. Royal, M. J. and Nimo, N. M. (1973). Big methanol plans offer cheaper LNG alternative, *Oil Gas J.*, **71**, 52–55.
18. Schulten, R., Kugeler, K. and Barnert, H. (1974). L'Energie nucléaire pour l'approvisionnement en énergie et en matières premières, *Rev. Française de l'Energie*, **265**, 253–261.
19. Soedjant, P. and Schaffert, F. W. (1973). Transporting Gas—LNG *v.* methanol, *Oil Gas J.*, **71**(24), 88.
20. Speich, P. (1974). Kohlevergasung—ein Beitrag zur Energie und Rohstoffversorgung, *Energiewissenschaftliche Tagesfragen*, **24**(7), 323–337.
21. Jamieson, J. K. (1973). Looking ahead to energy sources of tomorrow, *Exxon Marketing Exchange*, No. 6, 13.

Appendix A

Composition and Properties of SNG Process Gases

Composition and Properties of SNG Process Gases

	CRG		Gasynthan (b)	JGC/MRG (c)	GRH (naphtha) (d)	FBH (crude) (e)	Lurgi (coal) (f)
	Double methanation (a)	Hydrogasification (a)					
Composition, % vol							
CH_4	98.5	98.30	98.1	98.0	42.9	79.5	95.1
C_2H_6	—	—	—	—	24.3	8.7	0.7
H_2	0.9	1.05	0.7	1.5	25.8	10.0	0.1
CO	0.1	0.15	0.1	0.01	3.2	1.5	1.0
CO_2	0.5	0.50	1.1	0.49	0.2	0.2	
Propane	—						
Heavy hydrocarbons					3.6		0.6
Nitrogen	—				—	0.1	2.5
Oxygen						—	
Heating value, Gross kcal/Nm³	9548	9548	9531	9500	10006	9519	9099
Gross Btu/scf, 15° C, dry	1000	1000	998.2	995	1048	997	953
Relative density, 15° C, 1 atm (air = 1)	0.557	0.555	0.560	0.553	0.605	0.557	0.558
Wobbe Index Btu/scf, 15° C, dry	1342	1342	1331.1	1338	1347	1336	1289
Flame speed factor Weaver	14.2	14.4	14.2	13.9	21.8	36.9	14.9

(a) Chapter 6, Table 6.1;   (b) Chapter 6, Table 6.2;   (c) Chapter 6, Table 6.3;   (d) Chapter 7, Table 7.1;   (e) British Gas Data Book. Volume 1, British Gas Corporation (1974);
(f) Chapter 9, Table 9.1.

Appendix B

Typical Flowsheet of an 'Integrated Refinery' (Simplified: No Purification Plant Shown)

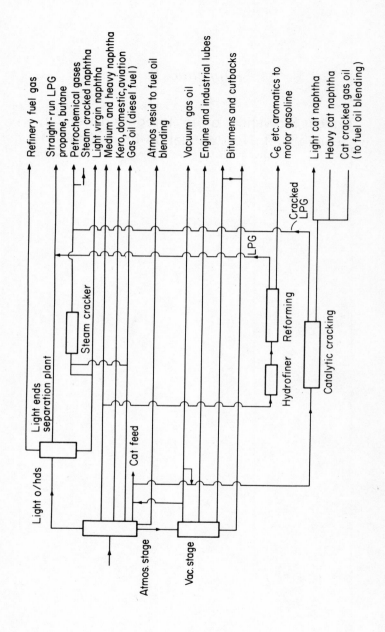

Appendix C

Economic Analysis of Industrial Processes

A. Investment

  (a) On site—all equipment purchases, on-site manufacture, erection and construction, testing, insulation, painting within battery limits.

      Includes:

Process units	Compressors
On-site storage	Services (distribution)
Buildings	Civil Engineering
Pumps	Instrumentation
Chemicals (initial charges)	
Catalysts (initial fill)	

  (b) Off site—all equipment purchases and manufacture at the plant to provide services and facilities required for production.

      Includes:

Steam raising	Sewers
Boiler feed preparation	Roads and fences
Cooling water facilities	Materials storage
Power generation	Offices
Compressed air	Social facilities
Inert gas	Environmental work

  (c) Engineering—all services by contractors, own staff and consultants connected with planning, design, construction and commissioning.

      Includes:

Research, development	Plant design
General planning	Ordering—expediting
Engineering studies	Construction supervision
Economic studies	Commissioning
Process design	Start-up

  (d) Working Capital—cost of plant operation including raw materials between dates of payments due and receipt of payment for product.

      Includes:

Stock of raw materials	Labour costs
Debts by customers	Other plant expenses
—less suppliers credit	Cost of finance

  (e) Interest during Construction

B. Operating Economics
    Expenses:

Capital charges:	Interest on borrowings	Maintenance
	Depreciation allowance	Repairs
	Process royalties	Loan repayments
Labour charges:	Operating labour	Overheads
	Supervision	
Materials cost:	Feedstocks	Aux. materials
	Chemicals	Catalysts
	Utilities (steam, water, power, air)	
Other expenses:	Insurance	Rentals
	Purchased services	Administration
Taxes:	Local	Pollution
	Property	

(b) Income:

Product sales:	Main product	Utility sales
	By-product credits	

C. Evaluation of Project
    Return on investment:

$$\text{Before tax} = \frac{(\text{Income} - \text{Expenses}) \times 100}{\text{Investment}} \%$$

$$\text{After tax} = \frac{(\text{Income} - \text{Expenses} - \text{Income Tax}) \times 100}{\text{Investment}} \%$$

Discounted cash flow return: If operating economics vary from year to year or if the project has a limited life it may be preferable to set up a cash flow table for the entire life of the project—including construction—and to discount all cash flows at increasing rates of interest until the total, the present value of the project, equals zero. The corresponding interest rate is the DCF return.

Appendix D

Calculation of Gas Service Cost for SNG Processes

Step 1   Calculate total annual fixed charges ($) inclusive of:
- (a) annual capital charges: 20% on capital to provide for repayment of capital etc.
- (b) annual operating charges: 6% on capital to cover operating, maintenance and overheads.
  Add (a) and (b).

Step 2   Calculate annual variable charges ($):
- (c) catalyst and chemicals, annual costs.
- (d) annual utility requirements.

Step 3   Allow annual credit for by-product sales or consumption (e):

Step 4   Calculate total annual charges (exclusive of feedstock): (a) + (b) + (c) + (d); subtract (e); divide by load factor (330 days, see Table 11.1) to obtain daily cost ($/stream day).

$$\$/sd = \frac{(a) + (b) + (c) + (d) - (e)}{330}$$

$$\cent/\text{million Btu} = \frac{(a) + (b) + (c) + (d) - (e) \times 10^8}{330 \times S \times CV}$$

where S is the plant output of SNG expressed in million cubic feet per day, and CV is the nominal calorific value of the SNG sent out (Btu/ft^3).

N.B.   1.   In many instances, the plant output is assumed to be 250 $\times 10^6 \ ft^3/day$ and the CV, 1000 Btu/ft^3 (see Table 11.1). In these cases, multiply $/sd by 0·0004 to obtain ¢/million Btu.

2.   Refer p. 208 (section 11.6) and Fig. 11.1 for calculations of total gas price inclusive of feedstock cost.

Appendix E

Important Conversion Factors (*SI Units)

Length	1 cm	$= 10\,\text{mm} = 0.01\,\text{m}^*$	
		$= 0.032\,8\,\text{ft}$	
		$= 0.393\,7\,\text{in}$	
	1 ft	$= 12\,\text{in} = 0.333\,\text{yd}$	
		$= 30.48\,\text{cm}$	
		$= 304.8\,\text{mm}$	
		$= 0.304\,8\,\text{m}$	
Area	1 cm^2	$= 100\,\text{mm}^2 = 10^{-4}\,\text{m}^{2*}$	
		$= 1.076 \times 10^{-3}\,\text{ft}^2$	
		$= 0.155\,\text{in}^2$	
	1 ft^2	$= 144\,\text{in}^2 = 0.111\,1\,\text{yd}^2$	
		$= 0.929\,03 \times 10^3\,\text{cm}^2$	
		$= 0.929\,03 \times 10^5\,\text{mm}^2$	
Volume	1 cm^3	$= 10^3\,\text{mm}^3 = 10^{-6}\,\text{m}^{3*}$	
		$= 35.315 \times 10^{-6}\,\text{ft}^3$	
		$= 6.102\,4 \times 10^{-2}\,\text{in}^3$	
	1 ft^3	$= 1\,728\,\text{in}^3 = 0.037\,04\,\text{yd}^3$	
		$= 28.316\,7 \times 10^3\,\text{cm}^3$	
		$= 28.316\,7 \times 10^6\,\text{mm}^3 = 0.028\,32\,\text{m}^3$	
also	1 UK gal	$= 1.201\,\text{US gal}$	$= 0.160\,55\,\text{ft}^3$
		$= 4.545\,96\,\text{l}$	
	1 US gal	$= 0.832\,67\,\text{UK gal}$	$= 0.133\,7\,\text{ft}^3$
		$= 3.785\,3\,\text{l}$	
	1 ft^3	$= 7.481\,\text{US gal}$	$= 6.24\,\text{UK gal}$
Density	1 g/cm^3	$= 1.000\,028\,\text{kg/l}$	$= 10^3\,\text{kg/m}^{3*}$
		$= 62.43\,\text{lb/ft}^3$	
	1 lb/ft^3	$= 0.133\,66\,\text{lb/US gal}$	$= 0.160\,53\,\text{lb/UK gal}$
		$= 0.016\,02\,\text{g/cm}^3$	
		$= 16.018\,5\,\text{kg/m}^{3*}$	

Pressure	1 lb/in² abs.	= 27·68 in w.c. = 2·307 ft w.c. = 2·036 7 in Hg = 51·715 mmHg (torr) = 0·070 306 kg/cm² = 0·068 027 atm = 0·068 948 bar = 0·689 48 N/cm²*

$1 \text{ lb/in}^2 \text{ abs.}$
= 27·68 in w.c. = 2·307 ft w.c.
= 2·036 7 in Hg = 51·715 mmHg (torr)
= 0·070 306 kg/cm²
= 0·068 027 atm
= 0·068 948 bar
= 0·689 48 N/cm²*

1 bar
= 1·019 72 kg/cm²   = 750·06 torr
= 0·986 92 atm   = 10 N/cm²*
= 14·50 lb/in² abs.
= 401·47 in w.c.   = 29·53 in Hg

Energy   1 Btu
= 2.93 × 10⁻⁴ kWh = 1 × 10⁻⁵ therm
= 3·93 × 10⁻⁴ hph
= 0·251 996 kcal
= 1·055 06 kJ*

1 kcal
= 1·162 8 × 10⁻³ kWh
= 3·968 3 Btu
= 39·68 × 10⁻⁶ therm
= 4·186 8 kJ*

Gas volume 1 scf (Standard cubic foot) (60 °F, 30 in, sat.) = 0·026 41 Nm³
(15 °C, 760 mmHg, dry)

1 Nm³   = 37·875 scf

Gas density 1 lb/scf   = 17·187 kg/Nm³*
1 kg/Nm³   = 0·058 18 lb/scf

Gas calorific 1 Btu/scf
value
= 1·018 Btu/ft³ (60 °F, 14·7 lb/in² abs., dry)
= 9·548 kcal/Nm³ (0 °C, 760 mmHg, dry)
= 39·978 kJ/Nm³*

1 kcal/Nm³
= 0·948 kcal/m³ (15 °C, 760 mmHg, dry)
= 0·104 73 Btu/scf
= 0·106 61 Btu/ft³ (60 °F, 14·7 lb/in² abs., dry)
= 4·187 kJ/Nm³*

Source: Williams, A. F. and Lom, W. L. (1974). *LPG Guide to Nature, Properties and Applications*, Ellis Horwood Ltd, Chichester.

Index